THE IDLE BEEKEEPER

THE IDLE

BEEKEEPER

The Low–Effort, Natural Way to Raise Bees

BILL ANDERSON

Abrams Press, New York

ABRAMS The Art of Books
195 Broadway, New York, NY 10007
abramsbooks.com

To Lesley, queen to my drone and our brood

"The keeping of bees is like the direction of sunbeams."
HENRY DAVID THOREAU

CONTENTS

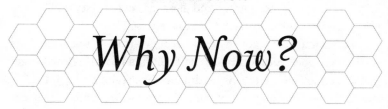

Why Now?

Almost everyone now knows how much we all depend on bees. In 1958 the Chinese found out the hard way in a cautionary tale of biblical proportions. Agricultural targets were not being met, and Chairman Mao decided that sparrows were responsible for eating intolerable amounts of the people's rice crops. He orchestrated a hugely successful campaign to eradicate them. The sparrows. The whole population was encouraged to get involved. Good citizens used guns and catapults to shoot the birds. They banged pots and pans, beat drums: anything to scare the sparrows from landing, forcing them to keep flying until they fell from the sky, dead from exhaustion. Nests were destroyed, eggs smashed, nestlings killed, and the sparrows were driven to near-extinction.

But with the sparrows gone, the insects they ate flourished. Plagues of locusts wreaked even greater devastation on the crops. Chinese scientists then actually looked inside dead sparrows' stomachs and discovered 25 percent human crops and 75 percent insects. Oops. So now the insects got it. Patchy and panicky misuse of pesticides like DDT not only wiped out the locusts, it killed all the pollinators as well. The bees weren't the only innocent victims of these acts of devastating human ignorance: the ensuing ecological disaster exacerbated the Great Chinese Famine in which at least twenty million people died of starvation.

In 2012 a study commissioned by Friends of the Earth estimated that if the United Kingdom couldn't rely on insect pollinators and had to do the job ourselves by hand, it would cost the economy

$2.3 billion a year—for a green and pleasant land of sixty million people that doesn't grow enough food to feed itself.

That same year I got my first bees. But it wasn't as a result of terrifying, apocalyptic statistics. I'd been in a beautiful English garden directing *Lewis,* a television drama whose pilot I'd "helmed," in the oddly nautical language of our industry. We were filming a scene that involved a fictional family enjoying a relaxing lunch alfresco. The laws of filmmaking usually insist upon hurricanes or snow on these occasions, but it appeared we had been given special dispensation: it was merely unseasonably cold. Actors were huddled like convicts in huge quilted thermal coats sprayed with stenciled coat hanger symbols—wardrobe department graffiti to discourage any thoughts of theft or style. Inelegant feather puffas were abruptly plucked away at the last moment before shooting the illusion of a midsummer, hypothermia-free take.

I was talking to the cameraman about how we might make the next freezing shot look warm and golden, and though he seemed to be listening intently, he began to slowly lower himself down to the lawn where we were standing and surreptitiously picked something up off the ground. Dropped litter I assumed, and carried on. But then, without taking his eyes off me, he tried to pretend that he wasn't gently blowing into the fist he'd made around the litter. Mysterious. He appeared to be still diligently listening to what I was saying, but I wasn't: I noticed he'd put his other hand into the pocket of his down jacket, a much more efficient way to keep it warm, so what was with the blowing?

Like a schoolboy caught stealing in a sweet shop, Paul slowly unfurled his fingers to reveal an insect lying on its back in the palm of his hand. Six legs skywards.

"Looks like a dead bee," I offered, none the wiser.

"Well, it might be . . ." Paul looked intently at the bee.

I didn't have to: "It's definitely a bee."

"Yes. But it might not be dead. If they stop moving for too long on a cold day, they can cool down so much their flight muscles stop working. Then they're stuck outside and they'll cool down even more and die. But sometimes, if you warm them up a little . . ."

. . . Two of the six legs twitched . . .

". . . just get them going again . . ."

. . . The dead bee flipped over off its back and started crawling on Paul's warming palm . . .

". . . they can fly back to the hive where there's warmth and food . . ."

. . . Lift off! I watched Paul's gaze following the bee as it flew off purposefully into the heavens. I'd just witnessed a resurrection.

"Not at all! I just gave her a helping hand," he said, but his eyes were twinkling with joy.

French philosopher Albert Camus said, "Life is a sum of all your choices." As a storyteller I spend most of my time interrogating characters' choices with the question "Why now?" It's the crux of dramatic narrative. All of us are all too aware that we often avoid making those significant life choices that will change everything: we know they will define us, determine our future, but how to get them right? The circumstances that demand such a choice may have been pressing for years, so when we finally commit and decide to act, what happened? What was the trigger? Why now?

I later found out that Paul, the cameraman shooting my drama, had another life: as Bond . . . Paul Bond, he was secretly a World Champion beekeeper. Even later I realized my choice to keep bees had secretly been made the moment that bee flew from his hand.

Right now you might like to consider why you've chosen to be reading this book at this particular moment.

The writing of it goes back to another, altogether grander English garden where the annual Port Eliot Festival beautifully meanders by a tidal estuary on the south coast. In the summer sunshine on the riverbank lawn I came across not a catatonic bee but a large tent: "The Idler Academy of Philosophy, Husbandry, and Merriment." The use of that last word outside the context of Christmas or alcohol brought a smile to my face and drew me inside. It was love at first sight. The Idlers' motto is *Libertas per cultum* and as well as teaching you Latin so you can knowingly pursue "freedom through culture," philosophy, astronomy, calligraphy, music, business skills, English grammar, ukulele, public speaking, singing, drawing, self-defense, taxidermy, harmonica, and many other subjects were also on the Idle menu along with much hilarity and merriment. But not Idle beekeeping.

The Idlers also publish a bimonthly magazine whose intention is to "return dignity to the art of loafing, to make idling into something to aspire towards rather than reject." Not out of sloth, but in a considered, slightly unorthodox, very low maintenance way, this is how I try to keep bees because I believe they know how to do it better than I do.

After the briefest conversation—"Can you write?"—I became the regular beekeeping columnist for the Idler magazine. Most of the participants in ensuing Idle beekeeping workshops assumed these gatherings where I taught the little I know were at least partly a ruse to sell copies of my book, but surprisingly not all of them were relieved to discover no such book existed.

In our biosphere there are outstanding individuals who have given their lives to studying and keeping bees either in an environment of dedicated peer-reviewed academic rigor or through decades of hands-on experience, often both. I am neither. I'm not even a shining example of Idleness. But I have spent many years telling stories for a living. Fabricating a tissue of lies into a believable semblance of truth is only possible through empathy, on the part of both the storyteller and their audience. The basis of the contract we make with each other is an agreement to empathetically work together to viscerally imagine and virtually experience the lives of others.

I wondered if the empathy that can allow us to understand and care for our fellow humans' dramas could allow us to do the same for the bees. But isn't that a bit soft and woolly? Where's the rigor? I come from Scotland, a country where "Physics" is still called "Natural Philosophy," squarely defining the study of all matter and energy in our universe as an act of human contemplation. This, too, is not possible without empathy. Nothing worthwhile is.

This book is going to show you why keeping bees with the minimum of intervention is positively good, and how to do it responsibly with minimum effort.

1

Idle Isn't Lazy...

. . . like a honey bee isn't a wasp. Superficially there are many similarities: size, shape, color, buzzing about, stinging. But although carnivorous wasps pollinate flowers, they don't provide honey. Vegetarian honey bees do.

And while both idle and lazy people seem to be similarly workshy, it's what they do with the time they're not working that makes the difference. Truly lazy people rarely cultivate themselves or the world around them. Idlers try to do both by cannily spending as little time as possible in drudgery so they can invest the maximum doing things that interest them, that make them grow.

The current rate of exchange offered by idle beekeeping is this: to get honey, wax, mead, and a share in the welfare of the bee population and the planet, you need to commit to attending your idle beehive on two occasions per year. That's not two days of work. It may be as little as a couple of hours, twice a year. You are of course entitled to spend as long as you like at the hive, entranced by the comings and goings of the bees, but this is idle contemplation, not laziness, and it's not compulsory even when it becomes compelling.

If having only two dates in your annual diary seems too good to be true, we used to get away with less. As beekeepers we effortlessly achieved a pinnacle of idleness for fifteen million years by employing the best and most basic strategy of work avoidance: don't turn up.

For fifteen thousand millennia before we took on the challenge of putting the Sapiens into Homo, honey bees were flying through the

OPPOSITE *Cave painting of human climbing vines to steal honey 6000 BC* air, navigating with astonishing accuracy to pollinate plants as they gathered food to raise their young in structures they constructed with mathematical precision. They worked out what they needed to do and they got on with it, without any help from us whatsoever, while we were still enraptured by our discovery of the amazing things you could do with the sharp edge of a bit of flint and opposing thumbs.

But before we were even standing upright on our two hind legs, we must have salivated at the smell of warm honey mellifluously floating down from high up in the tree canopy. And eventually connected the converging flight paths of bees as they congregated at their sweet source: a cavity in the trunk of a tree.

That was the moment we turned up at the bees' doorstep, and things got a lot less idle.

So overpoweringly delicious was the smell of that honey we risked climbing to great heights and braved the intense pain of hundreds of stings to steal the sweet treasure of the bees. When we mastered fire, smoke helped us confuse the bees and mitigate the stings. When we mastered tools, we were able to hack away at the tree and expand the entrance to the cavity so we could extract every last drop. We weren't beekeepers yet—we were bee killers. It wasn't just the honey we stole: the baby bees, larvae carefully nursed in their individual cells of the honeycomb, are an excellent source of protein and fat. We devoured them alive. Very few colonies of bees would have survived this onslaught, but we were hunter-gatherers of no fixed abode, took what we wanted, and moved on: pillage, not tillage.

Then around ten thousand years ago the idea caught on that instead of roaming out into the world to feed ourselves, we could make the world of food come to us. The revolutionary concept of agriculture and home. We began to settle down, and bees became a wild part of our farming. As with all the plants and animals we were cultivating, we would try to re-create their natural environment on our doorstep.

So we made containers with small, defendable entrances similar to the cavities where we'd seen bees living in trees. And when a swarm issued from one of those cavities and formed into a football-sized cluster of bees hanging from a nearby branch while it decided where to go and make its new home, we would come along and help it make up its mind by dislodging it into our container and taking it back to ours.

But from the get-go this was a process of enticement, not incarceration—to this day humans haven't devised a hive system with a lock and key. Unlike other animals we cage or fence in, the hive entrance has to be open for blossom-grazing bees to roam the airways as they please. They are wild like The Clash, and "Should I stay or should I go?" is the royal prerogative of their queen.

But she doesn't require any fancy abode—merely what her lineage of fifteen million years had become accustomed to. And the foundations of those palatial tree cavities that primitive beekeepers were trying to copy had been hollowed out by fungus. Those polypore spores that managed to penetrate the tree's defenses and start rotting down the wood fibers weren't guided by set squares, protractors, or spirit levels. Like us they were driven by appetite and organically munched away with the minimum of effort. So no straight lines or right angles for bee palace providers to pursue.

One of our more popular and enduring designs is the skep. Its familiar dome shape, made from a coil of twisted tall grass stitched together with the split stems of willow or bramble, is easy to make, light, and strong enough to take the weight of a man standing on it.

Inside, the bees would treat it in exactly the same way as an empty tree cavity they'd found: hanging from the top, they'd build pendulous, parallel combs of wax that they'd fill with babies and honey.

But we didn't make reusable skeps to save the bees, we made them to save us the work of finding wild colonies, then climbing and hacking away at trees. When it came to harvesting the honey, we still killed all the bees in our skeps to prize them away from it. The forest was still our near neighbor, and it was a regenerating source of swarms, so we didn't have to worry too much about where our replacement bees would come from. Indeed our skeps would themselves issue swarms that, with a bit of luck, we could capture in another skep.

We had become swarm-hunters-and-gatherers. But those clustering balls of adult bees looking for a new home contain no available honeycomb and no babies—very little nutrition for us, and all of it capable of flying away. So swarms had been of little interest until we became able to coax them into places where they would make babies and store honey. And then we started to see them as deferred deliciousness.

For thousands of years, whether in skeps, clay pots, logs, or anything else that we could fit bees into, we carried on killing bees to enjoy their honey.

Then on October 25, 1852, the carpenter's set square fought its corner and found the right angle to intersect with the world of bee-keeping: U.S. Patent No. 9300 was granted to the Reverend Lorenzo Langstroth for the design of a new beehive. And that's when things started to get precise.

Gone was the near-enough-is-good-enough of hand-twisted grass lashed together by eye. Langstroth's wooden hive was like a mini–chest of drawers lying on its back. With eight or more drawers that not only had to fit tightly together, they all had to open and close smoothly.

You'd need to be a carpenter bordering on cabinetmaker to have the level of woodworking skill to construct this hive. And yet

Langstroth's design now underpins the vast majority of beehives on our planet. Its global success is down to its ability to allow the bee-keeper to remove honey from the hive without killing the bees. And to extract that honey in industrial quantities.

Langstroth's hive neatly confined the bees' free-form honey-comb building into rectangular wooden frames, like drawers without a bottom. And these frames can be easily removed with their contents intact. Honey from a filing cabinet? It isn't quite that simple, because producing honey isn't the bees' primary purpose: it's having babies. They're all laid by the one queen bee living in the hive, and at the height of the summer she can deliver up to two thousand kids every twenty-four hours.

In my entire life I've had three. Admittedly the queen bee can have upward of twenty thousand family members actively helping raise hers, but there are some things we do have in common, and one of them is the way kids take over your home. Not in a directly tyran-nical way—they didn't ask to be born, as they will later tell you—but through the requirements and logistics their rearing demands. From the vomit stains on your shoulders that inspire you to ensure that every room in your house is filled with as many muslin squares as an artisan cheesery, to the ubiquitous scattering of toys that might briefly distract an inconsolably crying baby but will definitely hob-ble the sleepy, barefoot lullaby singer, needs-must rapidly takes over: diapers are never where they're supposed to be, they're where you last used them; food could be literally anywhere, from rusks in your undergarments drawer to bananas in your letterbox; and empty, unwashed bottles are usually lying where sleep finally released them from the gums of your progeny. Even when your children have grown into the bedrooms you've laboriously provided for them, the idea of any child-free space in your home is a delusional parental fantasy.

Likewise the beehive: dismiss any hopes of simply sliding out a frame of honeycomb that doesn't come with a brood of larvae wrig-gling in the middle of it, surrounded by baby food and a lot of pro-tective nurse bees.

So Langstroth, who also had three children, contrived a child-free space in his hive. Between two of the chest-of-drawer hive boxes there is a barrier full of holes of a very particular size. It's called a queen excluder:

The queen bee is larger than all the worker bees, and she is too big to pass through the holes in the queen excluder, so she's stuck in the bottom box, where she lays all the eggs. The smaller worker bees can easily crawl through the holes and move freely into the upper box, the child-free space, where they exclusively store the honey.

Langstroth's hive separates the babies from the honey in a way that allows the beekeeper to take that honey without disturbing the kids.

Top drawer. For the beekeeper.

Murder commuted to burglary from the bees' point of view. But at least now we qualified as true beekeepers: at the most basic level, we deliberately kept our bees alive, and we started to really make them earn their keep.

Langstroth's patent came seventy years after patents by James Watt for his steam engine and James Hargreaves for his textile spinning jenny, inventions that combined to create the first factories of the Industrial Revolution. So the universal benefit of human intervention to maximize productivity and efficiency was by now an established creed. And because Langstroth's frames allowed combs to be individually pulled out, forensically examined, and then replaced, the inner workings of the living hive could be observed. Through the eyes of Industrial Revolutionaries:

This representation of British society—nine layers of classes and trades, with the bank, armed services, and volunteers at the foundation and the monarch at the top—was etched by George Cruikshank, one of Charles Dickens's illustrators, fifteen years after Langstroth's invention. In 1867 Cruikshank published it to make the political point that everything was buzzing along beautifully and there was no need to give the working man the vote.

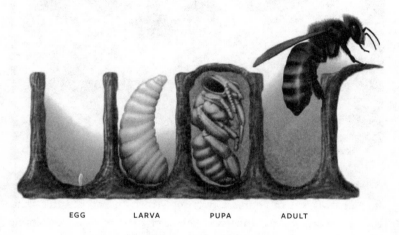

EGG   LARVA   PUPA   ADULT

The Life Cycle of a Bee

The honey bee colony consists of around fifty thousand bees, the vast majority of them female worker bees whose short careers develop through a progression of jobs. After three weeks as larvae then pupae, they hatch as adults. For the next three weeks inside the hive they will perform a plethora of distinct tasks including cleaning, nursing, housekeeping, undertaking, and guarding the entrance before a final three weeks of flying outside the hive to collect food. A much smaller number of bigger, male bees—called "drones" because their larger wings give their buzz a lower, louder tone—have only one primary job: inseminator. And an even larger single queen bee, usually found at the bottom of the hive, not the top, gives birth to all the children.

Our perception of the industrious bee society as a role model for our own was of course a two-way street. In Langstroth's time the buzzword for the hive was "hierarchy": nowadays "collaborative superorganism" might be a better fit—the bees haven't changed much in all that time, just our projections of ourselves.

But because Langstroth allowed us to get under the bonnet of the beehive, we became able to start tinkering about, or "improving." And our goal was sweet and clear: more honey. So we manipulated the environment for the bees inside the hive to exploit their behavior in ways that maximize honey production. We did everything we could to turbocharge that one aspect of the bees' lives. But as any car mechanic will tell you, whilst it is possible to turbocharge any combustion engine to deliver huge amounts of power, if the components of that engine were never designed to cope with all the additional stresses incurred by the increased workload, it won't last long. It will quickly burn out or break down.

Have we brought our bees to this point?

Imagine life as a bee in California this spring—not glamorously buzzing around the flowers at the foot of the "Hollywood" sign but approaching the one million acres of Californian almond trees coming into bloom in one of 120 hives on the back of a truck after traveling two thousand bumpy miles.

Living in trees never felt like this, even during earthquakes. But once the forklift in the almond orchard has unloaded your hive, you fly out into a world of blossom. The soil and climate here make it the perfect place to grow almond trees—80 percent of all the almonds eaten in the world call this home—and almost nothing else is grown. Every intensive inch is almond. And every almond needs a bee to pollinate it into existence. For miles in every blooming direction it's All-You-Can-Eat nectar and pollen for the bees, and every year thirty billion of them are trucked in from all over the United States for this feast.

But the food is just one flavor, and it lasts just a few weeks.

Each individual flower is available for pollination for only five days. Once that blossom is fertilized by bees gathering pollen, the almond tree quickly switches its floral energy and resources to making almonds, and there is nothing for the bees to eat. No nectar or pollen for miles. The thirty billion bees, who could be forgiven for thinking they'd only just got here, must now be urgently trucked out again to somewhere else with flowers or they will starve to death in a food desert.

Like a touring stadium rock band, but on a much bigger scale, the bees are forklifted back onto giant trucks that head out on the highway and race them to their next exclusive gigs: cherries, apples, plums, avocados, pumpkins, blueberries, cranberries, sunflowers, and vegetables. It's estimated that one mouthful in three of everything we eat is dependent on insect pollination. And honey bees are the rock stars of pollinating insects, driven by diesel and heavy metal thunder.

But is the on-tour, rock star lifestyle one you would wish on your family?

Parents of small children know that when they start school they bring home all kinds of diseases that are generously shared in the classroom. Imagine how stressful a National Infant School Camping Month would be if entire families had to travel hundreds of miles to congregate in mass camps of similarly dislocated people—even if the food was good.

The thirty billion bees that traverse the United States every year travel on frames in hives still based on Langstroth's design. But the honey that design was intended to maximize has become less important to the road warrior industrial beekeepers that collect it. Recently the commercial value of the pollination services their bees provide has become greater than the income from their honey. In cash terms, the honey is not so sweet.

So is Langstroth's honey-maximizing design still fit for purpose on the spreadsheets of all commercial beekeepers?

Small-scale beekeepers like me are not generating income from honey to feed their families, let alone satisfy shareholders. For us "hobby" beekeepers, many with a handful of hives or even just one, how important is maximum honey production? Does it trump everything else? And yet most of us still use the same hives and methods as the industrial giants. These hives have become synonymous with beekeeping. As if there is no other way.

But another example of small-scale animal husbandry is keeping a few hens to provide delicious fresh eggs for your family. Do the people who keep poultry in their back gardens always visit the nearest industrial sheds of battery hens and copy the farmers' methods of egg mass-production? And buy the equipment they use? And treat their hens with the drugs the farmers have to administer to keep theirs alive?

In answer to the perpetual conundrum "Which came first, chicken or egg?," the priority of commercial egg producers has to be the egg. The profit margin on the sale of every egg they produce is the basis of their business, and from that they work back to the hens that lay them, minimizing costs all the way.

As someone who consumes two raw eggs every day, for me the chicken always comes first. We can't help the egg, once laid, do anything to make itself more healthy or nutritious. But the hen can be helped to do both for her eggs: the better we nurture and nourish her, the better the eggs she lays.

Commercial beekeepers have to start with the profit margin on the sale of their honey and work back to the bees, maximizing productivity and minimizing costs along the way. That's business, and it allows millions of us to enjoy honey on demand, and at what appears to be at first sight, reasonable cost. A jar of honey on our supermarket shelves doesn't cost much more than a jar of jam, so they feel like a fairly equivalent treat. But the factories that boil the sugar and the fruit into jam shoulder no responsibility for pollinating into existence every third mouthful we eat. The bees that make the honey do.

So as we begin to understand that the health of the bees may be more important than the short-term wealth we accrue from their honey, maybe it's time to unbolt the turbocharger and get back to the trees. Not as a retreat from our modern industrial life, but for inspiration: trees successfully gave a healthy home to bees for millions of years when our workload was zero. But so was our honey harvest.

Can we get close to the giddy heights of idleness we enjoyed before we turned up at the bees' doorsteps and also enjoy a supply of murder-free honey?

There is a hive system that closely imitates the tree cavity, barely interferes with the bees' natural behavior, but allows us to gently remove their surplus honey with an absolute minimum of effort and time. And if I can keep bees this way, anyone can.

2

Ideal Homes

You may already have a place in your heart for your bees, but this doesn't automatically translate into desirable accommodation. Whatever your species, the same three golden rules for choosing your ideal home seem to apply—location, location, location—so where you plan to put your hive is crucial for both you and your bees.

Whilst we humans may prioritize easily accessible supermarkets, farmers markets, or organic whole-food temples, all the bees require is nearby flowers that we haven't made poisonous. Not an unreasonably high expectation to have of your neighbors, but one we disappoint all too frequently. And though even the wildflowers that grow in the margins of industrial agriculture are often tainted by the overspray of pesticides harmful to bees, all is not gloom: in towns and cities millions of gardeners deliberately plant for blossoms from March to November. From window boxes to public parks, we've intervened to generate the sight and smell of beautiful flowers that profoundly gladden our hearts, and incidentally provide bees with a varied, reliable, and extended supply of food—largely unsprayed and non-GMO.

Most of the locations we plan for our bees will fall somewhere between genetically modified super-prairie and unspoiled Garden of Eden, but there's often a concern that even if there is quality local food, will there be enough?

I was visiting a friend on the Isle of Harris in the Outer Hebrides who breeds "black" bees indigenous to the UK. The purity of the genetic line of Gavin's bees depends on the colonies being kept away

OPPOSITE *The Golden Hive*

from inseminating "outsiders." As I drove for hours through sparsely inhabited countryside, I was struck not so much by Gavin's splendid isolation but by the almost complete absence of vegetation—the strikingly beautiful rocks, occasionally punctuated by tiny pools of rainwater, initially gave the impression of a quarry. Miles of granite later, I began to feel that I could be on another planet where "green" had yet to evolve. But in the middle of this volcanic vastness were prospering colonies of bees that were finding plenty of food, because they knew what to look for: tiny pink flowers on leafless stalks that seem to float above green cushions of thrift eking out an existence in hairline cracks, and wild honeysuckle erupting from unassuming gaps in the rocks to release a cool flow of nectar and pollen from a cascade of golden blooms.

Within easy reach of my own bees in London, there is a wildflower area preserved in Wormwood Scrubs, a public park that adjoins the famous prison. When I walk our dog through tens of thousands of wild blooms, I frequently struggle to use up the fingers of one hand to count the honey bees feeding, and I've never seen them queuing up or fighting to gain access to a flower—even when the bees do turn up in huge numbers. I've seen dogs occasionally disputing ownership of a muddy, saliva-drenched old tennis ball, but I've never seen honey bees kicking some other pollinator off a blossom. Unless you're planning intensive bee farming, there seems to be plenty to go round.

So if your neighborhood is likely to be conducive, what about your neighbors? You are going to be introducing a wild animal into their lives as well as your own: if next door decided to keep a tiger as a pet, you might want to have a chat with them about the state of the garden fence. I dropped my nearest neighbors a note explaining my plans before I got my first bees. Included was a contact phone number and a picture of a honey bee—it's surprising how many people think all bees are bumblebees—they're not. Many are struck by how much honey bees look like browner versions of wasps. This is a distinction that needs to be made clear—though there is a lot of

WASP   HONEY BEE   BUMBLEBEE

sympathy around for the struggling honey bee, I wouldn't like to be a wasp-keeper trying to persuade my neighbors of the merits of wasps.

Everyone seems to know that bees die if they sting you, but many harbor the suspicion that wasps sting you just for the hell of it, and might do it again for kicks. And again. Sweetly, people assume the honey bee knows that stinging you will be fatal and therefore considers that option very carefully, and only as a last resort like the leader of a world power with their finger on the nuclear button. In fact, although the honey bee will only use her sting defensively, it comes as a complete shock when she flies off to discover that she can't retract her sting from human skin and has just terminally evis- cerated herself—this never happened before when she was dealing with other insects, with whom she could use her sting again and again: as devastating as a spray of bullets from an Uzi . . .

I left any reference to automatic weapons out of the note to my neighbors, but there is a legitimate concern about insect stings caus- ing anaphylactic shock. And if you increase the local population of stinging insects, the risk of being stung increases, even for bee-lovers trying very hard to never inadvertently threaten any bees—everyone makes mistakes. But risk is different than fear. If neighbors are con- cerned that they and their children, pets, or livestock might have an extreme reaction to a bee sting, that is fear. Valid fear, but whose proportions may bear very little resemblance to the risk. Like you being justifiably afraid of the tiger you just imagined prowling next door—the huge one with the very big teeth that already had its eyes burning into you—even though it doesn't exist. Probably . . . But if you have neighbors with prior experience of allergic reaction to stings, then you really need to consider the risks carefully. Death from anaphylactic shock is as final as death by tiger, so you may have to consider placing your bees somewhere else, farther from any sensitive humans.

A couple of years after I installed my first bees on our roof, we were unusually away on holiday during the Notting Hill Carnival and returned to find that our teenage son had turned our entire basement into a community rehearsal space for local bands—carpets had been dragged out of nearby skips and thrown down on the floor and up the walls. Guitars and amplifiers vied for floorspace with the drum kit, a precarious keyboard, and slumbering musicians. All silently covered in party detritus. As they surfaced and began wearily clearing up, I heard them sniggering as they impersonated a furious neighbor who'd evidently come round at four a.m. and apoplectically demanded they stop playing.

I went round to add my apology. What seemed to annoy him most was they'd kept playing the same song. For hours. While I was in mea culpa mode I thought I might as well include the bees, and asked him how much of a nuisance *they'd* been over the last two years: "Oh, did you actually get any?" was his reply. He hadn't even noticed, so no palpable increase in his risk of getting stung, but by now there were four hives up on my roof—how could their inhabitants have escaped his attention?

Bees don't require you to seek planning permission to build a concrete runway on your property, but they do immediately bring a flight path whose arrivals and departures from the hive would leave a human air traffic controller breathless.

Airports have runways that take advantage of the prevailing winds to help aircraft fly. Sunlight isn't a factor in deciding the direction of the flight path—747s are equally comfortable taking off and landing in the dark. Bees, on the other hand, generate their own lift when flying, like helicopters. They don't need to exploit the breeze to get airborne, but they do need to know where they're going, and their guidance systems calibrate to the sun. Positioning your hive so its entrance faces the prevailing direction of the sun, south in the northern hemisphere, allows the bees to check their bearings as soon as they fly out.

Even when it's cloudy, bees' polarized eyesight allows them to see the position of the sun, but when it disappears behind our planet they are unable to navigate and return to the hive for the darkness of night. Light on the dawn horizon is their wake-up call, so an entrance including some eastern promise in its southerly aspect alerts them to the start of the working day as early as possible.

The vast majority of flying bees are collecting food, and most of them already know where they're going as they leave the hive—directions to the various locations of nourishing flowers are shared by those foragers that found them. Flying out to their nectar and pollen destinations, bees will fan out from the entrance and up into the air. They'll return by the same route. It's this busy cone of determined airborne bees that we need to position with care: stand behind the hive entrance and there's nothing going on; stand in front of it and at peak times five hundred bees are coming and going every minute—you and any children, pets, or neighbors will be very in the way.

Imagine an invisible, weightless foghorn with a mouth bigger than an adult. Attach it to the front of your hive, point it toward the sun, and see where it will fit into your life.

If available space falls a little short of your full foghorn, you can use or create an obstruction that will divert the bees upward sooner: a human-height hedge or a panel of trellis or fencing near the hive entrance will ensure that the bees' flight path doesn't disturb. They will patiently accept this interference as a landmark: millions of years living in trees in forests full of other trees has given bees a lot of experience negotiating permanent obstacles near takeoff and landing.

Because my hives are up on the roof of our house, their entrances are already higher than anywhere the public frequents. And because the foghorn of their conical flight path is upward as well as outward, my TV-antenna-and-chimney-dodging bees are even less likely to get in my neighbors' way.

Which probably explains why the teenage music pumping from the basement was perceived as the only nuisance.

But even height doesn't guarantee harmony in the hood. Whilst elevated bees may have spread out from the hive by the time they've flown down to plants growing at ground level, fanning out to disperse the "runway" effect, if there is something especially desirable right by the hive, lots of bees will find it.

A large concentration of blossom can be irresistible. This is of course a compliment to your neighbors' growing skills, and though they might share the bees' enthusiasm for the beautiful blooms of their prized shrub, this might not extend to having thousands of bees regularly flying in for a free meal without warning. Keener gardeners might be wooed by the fact that plants enthusiastically pollinated earlier in the season have to expend less energy over less time "putting out" flowers and nectar in the forlorn hope of seducing passing bees. Although the flowers may be beautiful, they're just a means to an end—and the quicker plants can get on with successfully reproducing themselves, the more sunlight they'll have available to make the energy they need to grow stronger, or have even more offspring, or both, before the onset of winter. Bees make plants healthier.

But if this prognosis isn't enough, it's worth reassuring your neighbors that bees visiting their flowers are much less likely to sting than the bees at the entrance of your hive. Even if the two are very close.

The reason behind this certainly worked for me. For years the thing that stopped me keeping bees was a highly developed fear of them. And all stingy, buzzy, danger-stripey flying things. Of course, this was more a reflexive fear than a rational one. In the style of Mr. Bean, or on a good day Jacques Tati, I would comically recoil from an approaching bee much, much faster than I could come up with a reason for doing so. Let alone a punchline. Running away just happened all by itself, and the bees seemed comfortable to see me go, without sharing any of my humiliation—those danger stripes and stings didn't evolve to encourage intimacy. So before I actually committed to getting my first bees I arranged for a close encounter—to see if I'd be able to resist any reflex that might compel me to jump off my roof.

A frame of honeycomb covered with thousands of bees was held inches from my face. A millisecond passed. I'd never been so close to so many bees for so long. The rational side of my brain was plunged into its greatest-ever struggle with my amygdala's instant calculation of the orders of magnitude of overwhelming fear that this insane bee exposure was going to unleash. But one second later, both cerebral combatants realized that the bees were still there. And so was I. Astonished. So astonished, I took the honeycomb in my hands and brought it even closer with even less fear. The torrent of bees I had feared would rush into my face and sting me, blind me, were carrying on about their business calmly. My rational side couldn't decipher the detail of what they were up to, but all my instincts could feel their level of concentration. They were brimming with purposeful intention, and that intention wasn't me, it was their kids. The business of raising children was clearly far more important and all-consuming than wasting time and effort pointlessly stinging anyone. They paid me no heed. But as a parent, I knew that if I posed a threat to their children I would instantly become the center of their attention and not in a good way.

So why would this reassure my neighbor with the luscious blooms? If he accidentally came between my bees and his plants, wouldn't he invoke their ire as surely as if he stood between that tiger and its prey?

Imagine you're in a supermarket standing in front of the preserves shelves agonizing whether to buy honey or jam. You definitely need something sweet to spread on your toast, but there's such a vast array, you're spoiled for choice. Someone glides past in front of you, effortlessly putting a jar of honey in their cart without breaking their stride. You might quietly envy their calm decisiveness, but do you rush after them, punch them on the nose, and take the jar of honey back? It's not yours!

Now imagine you're in your home having a family meal, and the imaginary gliding person from the supermarket steals into your house and tries to make off with that very expensive jar of artisan local raw honey you chose not only for its deliciousness but to ameliorate some family hay fever . . . . Now it's the intruder who's toast.

Your hive is full of baby bees that their parents will defend to the death. Your neighbor's blossom is a bee supermarket in the Garden of Eden—well stocked, but no price tags and no checkouts. The bees will not behave defensively over flowers they don't regard as "theirs."

The sugar content in the liquid nectar bees collect to make honey varies from plant to plant, day to day, even hour by hour, but usually the nectar is mostly water, and normally this is plenty for all their hydration needs, but there are times when more is required and they will fly to collect fresh. If you can call water from a nearby paddling pool full of children "fresh." If this aquatic garden fun is, as in most British summers, occasional and sporadic, bees may not discover it in significant numbers, let alone come to expect its water to be there when they need it. But if a source becomes more permanent, like a pond, swimming pool, or animal watering trough, it may become a feature of a hive's map of its world—and bees will return. Even when the pool that's been exclusively theirs for weeks suddenly fills up with kids.

Bees like to keep their feet dry when they're drinking—they've found it helps them drown less. None of the bees found floating in swimming pools planned to be there: until we turned up, they had safely assumed for millions of years that any expanse of water would reliably have naturally accessible banks or edges for them to stand on.

Providing the equivalent of a birdbath filled with semi-submerged pebbles *before* your bees discover the paddling pool next door will encourage your bees to drink responsibly, and help prevent nearby children from being stung—either because in fear they were trying to swat a flying bee away, or because out of kindness they were

trying to help a bee out of the water onto dry land and she mistook their giant attempt at help for an attack.

There is another vital element to placing your hive that I initially failed to consider, with catastrophic consequences: air.

Although my roof provides height it doesn't offer much flat space. We live in a Victorian terraced house, and the roof slopes down from the elevated chimney breasts we share with our neighbors on both sides. It's a "valley" construction allowing rainwater to run down two sides of sloping tiles to a horizontal channel that runs the middle of the length of the house. This horizontal channel narrows as it slightly falls to the drainpipe at one end, and there's only enough room to comfortably fit six hives up the other end.

I'd never thought I'd have more than a couple of hives, that there would always be plenty of space, but I discovered this limiting factor one gloriously fecund year when I was blessed with so many colonies of bees that there was no room in the valley for a seventh hive.

So I put the seventh hive up by the chimney pots. It blended in beautifully, and instead of polluting smoke curling out of the chimneys, now there were pollinating bees swirling to and fro at the entrance of the hive. On sunny summer evenings this hive was bathed in light long after the other six hives down in the valley had sunk into shade, its bees burnished into iridescence by the gently setting sun. I fell in love with it above all the others, this golden hive.

You can see my rooftop apiary from an elevated aboveground section of the Hammersmith and City line on the London Underground. Just a brief glimpse through trees as the train trundles past. Returning home I can never resist this different, secret view of my favorite hive . . . but like a beautiful smile destroyed by a knocked-out front tooth, mine was wiped off my face: an old gap between the chimneys had come back—the golden hive was gone!

I rushed home, clambered up onto the roof desperate to believe I'd somehow imagined it. But as I climbed up the sloping tiles to the chimneys where the golden hive had been, I saw where it had gone: it had fallen away from me onto the downward-sloping roof next door. I didn't need rocket science to work out what happened. In calculating the golden hive's flight path, I'd forgotten that airports don't just have runways: they have windsocks.

A gust of wind had tipped my golden hive over. And because my neighbor's roof was so far below the hive's base, it bounced on its head before its momentum flipped it over to crash down into the valley in an almost grotesquely perfect somersault. It was lying on its side, motionless. There was no inhalation or exhalation of bees at the hive entrance, no sign of life. The heavens began to weep.

It hadn't occurred to me that my hives in the valley were protected from the wind by the very chimney breasts I'd placed the golden hive atop—it couldn't have been more exposed—the dark side of all that sun.

Dazed with grief and remorse for causing the needless death of thousands of innocent bees, I set about removing the evidence from my neighbor's valley. I laid a ladder down the sloping pitch of his roof and climbed down to recover the fallen hive. It weighed the same as my sixteen-year-old son, but there appears to be no equivalent of a fireman's lift when it comes to bees, so I just clasped the hive to myself in a bear hug that was as emotional as it was inefficient.

Back muscles silently screaming, I started climbing the sloping rungs of the wet ladder as if they were stairs. Vertebrae were refusing to permit the mechanics of looking downward, but I overrode them to check my precarious footing and saw something that almost made me miss a rung—where the hive first hit the roof there was a gaping hole over a foot wide that allowed a clear view of the contents of my neighbor's attic, starting to glisten in the rain. Somehow I squashed the golden hive in with the others down in my valley and rushed back to the scene of my crime.

My emergency repair with breathable roofing membrane held for the six months before the roof was properly fixed. But after only three days a sight took my breath away—a bee, laden with pollen, flying into the entrance of the golden hive: and she wasn't an example of pilot error but part of precise, busy air traffic. These bees not only had somehow survived, it was business as usual!

This miraculous recovery obviously had nothing to do with me—I had provided these bees with a spectacularly unsafe home—its location, location, location was ideal, but I hadn't conducted even the most basic structural survey—the next essential box any prospective house hunter must check.

Beyond making sure that your hive isn't going to collapse, fall down, or be blown over, the bees spare us the need for surveyors' theodolites, plumb lines, and tape measures: although we might think we're providing spacious accommodation, all we're really giving them is an empty space in which they build their home all by themselves—not just the odd self-assembly kitchen unit, the entire edifice. We don't even need to provide construction materials—they're all flown in. We merely give them an empty box. But the least we can do is make sure that the space we're offering is as close as possible to that which the bees have been used to for millions of years. So let's take a closer look at those cavities in trees the bees choose to call home.

They tend to be tall and thin, not short and squat like most hives. And even this extraordinarily vivid attempt by the Natural History Museum to provide a view of a living hive inside a tree chooses a cavity that is still neither tall enough nor thin enough: This is not some arbitrary statement of fashionable taste following the current vogue for lanky supermodels. It's a mechanical observation about the "tree":

Look at the thinness of the walls of this cavity: they're around an inch thick. And now imagine that instead of a short, stumpy bit of tree trunk above it, there's fifteen or thirty feet of tree with hefty branches.

The weight of this tree above the cavity would be phenomenal, and though the groaning cavity walls in the picture might carry it for a second, the smallest puff of wind would create enough leverage at the top of the tree to snap the thin walls of the trunk at their weakest point—the cavity. Fallen tree, no cavity, homeless bees.

Professor Tom Seeley of Cornell University found that the thickness of the walls of tree cavities actually used by wild bees in his neighborhood ranged from an average of six inches to as much as twenty-nine inches. Much stronger than the one-inch walls in the picture, but much less room inside the tree:

Such a tree can sustainably accommodate such a cavity, and bees, for many years. Although six inches may be an average thickness of cavity wall, I like the look of the slightly better than average ten inches of wood surrounding the cavity revealed by the chainsaw in the photo. So let's emulate it in our choice of hive—crucially remembering that hives rarely have roots that anchor them to the ground like trees. Especially up on roofs.

But though the Natural History Museum's tree hive might not make it onto the catwalk of slender cavities, the wax honeycomb the bees have built inside it is textbook. Without a cabinetmaker's wooden frame in sight, the bees use their own expertise to fashion a precision structure. The first thing they do is form a chain gang—literally.

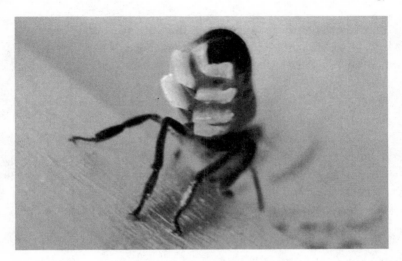

They then produce platelets of snow-white wax from glands in their abdomen.

And combine them together to add to the honeycomb under construction. The average time an individual bee spends on this task is thirty seconds. After that it moves out of the way to allow its wax glands time to replenish, and another bee takes its place to add its half-minute contribution.

There are no overseers referring to plans. With brains the size of a grain of salt, these bees are turning up at the wax face, instantly assessing the current state of construction, and building what needs to be built to move the project forward by thirty seconds. Because they can do this at whatever stage of construction they find on arrival, each bee must carry all the knowledge required to build entire honeycombs in her head.

And with extraordinary precision: the chain gang of construction bees uses gravity to ensure that the comb is built perfectly vertically—hanging freely in space, the bees are their own plumb line. We used to think bees created each of the perfect hexagonal cells not with protractors or laser guidance, but by initially building them as simple circles that fitted snugly around their bodies: once these wax tubes were completed, the bees would heat them briefly to make them fluid enough for the surface tension of surrounding cells to force them to flow into the familiar hexagons, which would set as they cooled. But we now know that during construction, bees never heat up the wax hot enough to allow such flow to occur, and we're back scratching our heads in mystified awe of how they achieve such precision.

Bees enhance the strengths of their hexagonal wax cells by building them in two layers, back to back in the vertical comb. The hexagonal patterns in the two layers of comb are offset from each other for structural support.

The ghostly outline of the Mercedes-Benz emblem seen through the bottom of each cell is structural reinforcement created by three cells meeting on the other side. Maximum strength for minimum wax. Homespun durch Technik, as Anglo-German engineers might say.

 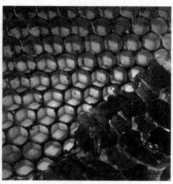

And in the almost complete darkness inside the hive, bees communicate by clinging to the comb and vibrating their bodies: the wax structure resonates so effectively at these frequencies that any bee walking on the comb can feel those coded vibrations and pick up the message being sent—beemail on the World Wax Web. It was this combination of Stradivarian streaming and structural strength that saved the lives of the bees in my golden hive: the force from the impact of the roofquake was directed through the strongest axis of the comb, minimizing damage and leaving plenty of sound comb to allow the vital communication that the repair teams needed to work together effectively.

Is it just me or is Langstroth's technical wizardry with a set square starting to lose a bit of its sparkle? I'm not denigrating chests of drawers—they're difficult to get right—but the honeycomb Langstroth's rectangular wooden frames contain is borderline miraculous.

It's highly unlikely that Langstroth understood what we now know about honeycomb as a means by which bees communicate, so he would have been unaware of any "firewall" effect his frames would have had on data exchange through what would be continuous comb in the wild. But if our imagination is now beggared by the little we *do* know of the level of communication the bees require to function as a complex social organism in colonies of up to sixty thousand, maybe we should be doing as little as possible to interfere with it. Do the nuances of a Stradivarius violin sound so good once it's been sawn into sections?

Back in 1917 French clergyman Émile Warré came up with a hive that combines the minimum interruption to the continuous comb the bees have relied on for millions of years with the ability to harvest their honey without killing them. After experimenting with 350 various types of hive, he settled on his design of the People's Hive detailed in his book *Beekeeping for All*. Warré—pronounced "War-Ray" like "x-ray"—not only wanted his People's Hive to be more like a tree cavity, he wanted it to be simple enough for anyone with the most basic woodworking skills to be able to make. He wanted beekeeping to be more popular and more democratic.

His brilliant hive is not the last word in idleness, but we can fix that without too much effort.

VENT

VENT

ROOF

QUILT

CLOTH

TOP BARS

BOX 1

TOP BARS

BOX 2

FLOOR

3

Follow the Honey

A Warré hive stands tall and proud. But in spite of all the thought and care that's gone into its design, it is still, fundamentally, a stack of empty boxes trying to pretend they're an injured tree. This injury would probably be decades old: a branch might have been snapped off in the high winds of a storm, exposing the sapwood beneath. The tree tried to repair its protective bark over this wound in two stages, much as we would attend to a cut in our skin: it sent antiseptic chemicals to infuse the surface of the exposed wood, and physically covered over the site of the injury by growing a bandage of scab-like wood callus. But if a tree isn't quick enough, fungi and bacteria will get in and start consuming it. And wherever there are thriving fungi and bacteria, there begins a food chain. From insects like tiny midge larvae grazing this microscopic mushroom garden, to beetles and ants that devour the softened wood the fungi and bacteria have predigested, all are delicious to the larger animals whose beaks and claws rip out the deadwood to get at them.

Some of these feeding sites are enlarged to become nesting sites for birds, and occasionally this evolution of excavation delivers a Goldilocks opportunity for a swarm of bees: an empty cavity that is just the right size for all the babies they're planning, with an entrance opening that's not too small to cope with their air traffic but not too big to defend.

And once a swarm's decided on their new home, they don't dither: as many as twenty-five thousand bees may move in within half an hour. It's unfurnished, and though they're all prepared to lit-

OPPOSITE *Exploded and clear*

erally "hang out" as a cluster clinging together around their queen at the top of the cavity, eggs don't have legs or arms to grab onto their neighbors, so the colony desperately needs somewhere safe to deposit its babies—hundreds of them, right now. Nursery building starts immediately, with wax. So urgent is the need for comb for kids that dandruff-sized flakes of snow-white wax start to build up outside the entrance to the tree cavity while the thousands of bees are still moving in—some of them literally can't contain the eight wax secreting glands in their abdomens until they get inside. But with greater thoroughness than any brush we might apply to the shoulders of our dark jackets, the bees will carefully collect every flake of this embarrassment of waxen riches and carry it inside to their construction site.

Their queen needs to lay hundreds of eggs every day to keep the population of this new colony viable. Each one in its own cell of honeycomb. The ability to make wax is only available to bees for three days around the sixth week of their nine-week life span. After that the specialist glands disappear like retracting undercarriage, and wax-makers become pilots. Younger bees must step up to take their place at the wax face, to provide for more babies, to grow into more younger bees, and this process must accelerate for the colony to increase and survive.

When we build a high-rise, we start from the bottom—lasers, spirit levels, plumb lines, and scaffolding guide our building up to the vertical. By starting at the top of their cavity, linking together in a chain that hangs hammock-like from the ceiling, bees require none of the external technology human builders rely upon to ensure that the comb they're building will be perpendicular. Collectively, the bees in those chains are both their own plumb line and scaffolding. Individual bees, hanging in space, know that all they have to do is build directly above their heads. Because unlike a human plasterer working on a ceiling, who's ensured that the scaffolding he stands on will protect him from the full effects of gravity, the comb-building bee is relying on the constant pull of the earth to make her position perfectly plumb. Once she and her interlinking sisters have built the first few levels of cells on the ceiling, the anchor points for the living chain of builders now become the comb itself, guiding its own construction directly downward to enable the bees to create a perfectly vertical comb face.

And though this uniform verticality is vital to the bees' survival—we'll see why in chapter 7—it's the downward direction of construction that's the basis of our ability to take honey from the Warré hive without killing the builders and all their family.

In the wild, a queen that just moved into an empty cavity with her swarming daughters is ready to lay eggs inside cells of comb the second they're built. They can't be ready soon enough. As these cells are filled with eggs, the queen follows hard on the heels of the construction teams, eager for the completion of any new nursery beds she can fill. Like an egg-laying surfer riding the crest of a wave of wax that's slowly moving downward, the queen delegates a crammed neonatal ward in her wake.

The bees will fill their cavity with comb, and depending on the size and shape of every last nook and cranny, there may be eight or more chain gangs building combs in parallel. These construction teams coordinate to leave just enough room between their combs for adult bees to have free access to feed the babies growing inside their protective cells—so two nurse bees working in the same position on neighboring combs will be touching, but not squeezing, each other's backs. Like "back-to-back" housing in cities, this technique maximizes accommodation in the available space.

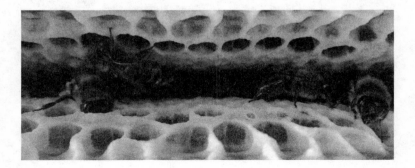

For nine days the firstborn are nourished by nurse bees, whilst beneath them their descending royal mother carries on laying as fast as she can. By the tenth day, the eggs that have grown into cute little wrigglers are fed up being larvae and are ready to pupate. They'll spin a cocoon just like a caterpillar on its way to becoming a butterfly, but because bee larvae spin theirs inside the protective wax hexagon of the comb, their silken solipsism becomes a hidden stage of their mysterious metamorphosis into an adult bee. The nurse bees tuck their charges in for the spin by sealing the entrance to each cell with a layer of wax that is different from the main structure of the comb—it's breathable so the adolescent pupaters can oxygenate their transformation. Twenty-one days after the first eggs were laid in the first cells, young adult bees will eat their way through this permeable wax cap and get to work, leaving an empty cell behind them.

This is the first moment in three weeks the queen has any choice as to where she lays her next egg: either she can lay it in the brand-new, snow-white, fresh wax cell by her feet at the bottom of the comb, or she can head upward, climbing around nurse bees feeding dozens of young larvae in open cells, then scaling scores of sealed, silken shape-shifters, to finally get to a vacant cell that looks and smells like an adolescent's spent his whole life in there without an en suite.

Even though housekeeping bees will give it a good scrape-out and coat it with disinfectant, this reconditioned cell is probably not going to be preferable to a new-build: would you rather lay your new baby down in a brand new white cotton diaper, or one outgrown by its previous owner who never washed it?

And it's not just a question of squeamishness. My family lives in a tall, thin, terraced house with many stairs. When we bring fresh food home, we carry it through the door at street level and into the kitchen at the bottom of the house. It feels more convenient than building a kitchen in the attic, carrying fresh food up sixty stairs every day, and then cooking and eating our family meals up there. We've not performed a detailed, scientific analysis of the cost benefit of relocating our kitchen to the top of the house, because it just doesn't seem to make any practical sense.

In most tree cavities chosen by wild nesting bees, the entrances are at the bottom, like in our house. And though our queen bee has the hypothetical choice to lay an egg in the cell that's just become available at the top of the cavity when her firstborn emerge into adulthood, nurse bees down at the bottom of the comb are feeding her babies with food that's not only nutritious, it's also nearby. Next to cells filled with babies are cells filled with baby food: pollen and nectar collected from blossoms are flown in and stored in the comb nearest the entrance, which also happens to be where the babies are. Not rocket science, but efficient: there's a nursery complete with its own sustaining infrastructure thriving down at the bottom of the comb, and the queen has the good sense to take advantage of it by preferring to lay eggs in the new comb being built just underneath it.

As increasing numbers of adult bees hatch above the nursery, their vacant cells become available for storage: not so much for baby food as for the more long-term needs of the whole colony. If we imagine the baby food storage near the baby bees is like fridges in our kitchens, the storage at the top of the comb is more like freezers in our garages—frozen food is not so immediately available or edible as fresh, but it keeps for much longer. The bees' equivalent of our frozen food is honey, and that is what they begin to store in vacant cells at the top of the hive.

So in the wild, the farther the industrious nursery moves downward into newly built comb, the bigger the larder full of honey in the older comb at the top of the tree cavity. Hold that sweet thought.

Hold it through the summer and into the autumn when fewer plants are putting out their blossoms for the bees to pollinate in exchange for food. Seeds and tubers are developing as plants now concentrate their energy on their young, and inside the hive the queen bee will be laying fewer eggs—she won't give birth to mouths her colony can't feed. Once winter shuts down their supply of fresh food from outside the tree cavity, a reduced population of bees will hole up inside and try to survive on the stored sunlight of honey until spring returns.

I would love to believe that bees can foresee the future. That they can accurately predict how cold and how long a winter will be before it begins its chill. That they could know in advance how much winter fuel they're going to need and save themselves the work of gathering more. That they could predict the numbers that will win the lottery. But there's no evidence that they can.

And we're no better. On Harris in the Outer Hebrides the flowering of the heather is a hugely significant annual source of food, and beekeepers whose scratch cards again failed to make them millionaires laboriously move their hives out onto the moors so the bees can gorge on this purple spread. But sometimes, just as the blossoms open, horizontal rain arrives and persists for three weeks, water-blasting all the anticipated pollen back into the earth. Bees and beekeeper can only watch from dry shelter.

When they are able to fly, bees tend to work at one speed—flat out—and as their population rises and falls with the availability of food, they simply store as much honey as they can.

A flower-rich summer allows an expanding population of bees to store large quantities of honey. If the winter that follows is not too severe, the bees may not need all that honey to see them through to fresh spring blossoms. There may even be a surplus.

Keep on holding that sweet thought . . . but let's see how the Warré hive can allow us to jump to its even sweeter conclusion.

It's not quite just a collection of empty boxes. Each box has some sticks. The fancy words for them are "top bars"—not to be confused with highly ranked watering holes—but they really are just sticks carefully spaced out at the top of each box for the bees to hang from and build comb.

A swarm moving into an empty Warré hive would find eight top bars on the ceiling conveniently spaced apart to roughly the distance the colony likes to construct combs to allow enough room for two bees to work back to back. Spookily those top bars are also the same width as a honeycomb. How accommodating. And the bees will start building comb downward from them in exactly the same way they would inside a tree. But once the combs have descended eight inches, bees working upside down at the wax face will feel something on their backs. Their first thought might be that they've hit the bottom of the tree cavity they think they're inside, but a quick look around reveals that it's only another eight sticks, also with gaps between them exactly wide enough for two bees working back to back.

They've reached the top bars of the next hive box, and there's loads more space below.

So the bees crawl through the gaps between those top bars that conveniently align with their beautifully vertical construction work, then hang from the underside of those top bars, and carry on building comb directly downward from there. Into the next box of the Warré hive. Until, eight inches later, there's that feeling on their backs again, and the bees have to work around the sticks and carry on down into the next box below.

If you were a bee walking down from the top of a fully occupied Warré hive, the comb would feel exactly like it had been built in a tree cavity: beautifully vertical and attached to the wooden sides. The only difference would be that, as on a tiled floor, though the surface you were walking on would be consistent, the pattern would be interrupted every eight inches . . . hexagon, hexagon, hexagon, wood, hexagon, hexagon . . .

Our purpose here is not to incorporate aesthetic marquetry into the dance floor of the comb. It's to take advantage of the bees' natural tendency to stock their larder of honey at the top of the hive.

The subtly interrupting top bars in the Warré hive allow its uppermost box, the one that's full of combs of surplus honey in the autumn, to be detached and removed, leaving the rest of the comb hanging from the top bars in the hive boxes below.

This process reduces the volume of the hive available to the bees—"What just happened to our cavity!" But we precisely compensate for that by adding an identical but empty box down at the bottom of the hive. Any hamster that's run on a wheel without getting anywhere would understand the basic principle of the hive. The bees keep building downward, but they never quite reach the bottom.

We've harvested a boxful of honey without killing bees or disturbing their main living area in a hive that allows them to build comb the way they would in the wild, gives them unrestricted freedom of movement, and ensures that they never run out of space. Sweet.

The Warré hive also requires very little intervention on the part of the beekeeper to perpetuate the cycle that allows us to harvest without harm: remove a box of surplus honey from the top in the autumn, and wait till spring before adding an empty box underneath. There are two good reasons for this delay.

The basic Warré hive consists of four hive boxes that are usually enough to give a colony plenty of room at the peak of its population in summer. As a rule of thumb, harvesting one box of surplus honey in the autumn would leave the bees comfortably off with three: a box and a half for stored honey—essential winter fuel—and a box and a half of comb for reduced nursery and accommodation.

Like canny aristocrats who might close down the east and west wings or mothball a few rooms of their stately homes to save the cost of heating them over winter, the bees will save quite a bit of honey staying snug and cozy in three boxes. If you add an empty fourth box immediately after the autumn harvest, not only would you be increasing the bees' heating burden for the coming months, as a beekeeper you would risk committing the greatest crime against idleness—knowingly doing unnecessary work. The three living hive boxes you would need to lift up so you could put an empty one underneath are currently laden with honey—maybe forty pounds of it to see the bees through the winter. They will be heavy. But if you wait till the spring, the bees will have consumed a significant amount of this honey, by which your load will be precisely lightened.

And on the subject of lighter loads, now is probably the time to be clear that if you choose to keep your bees in a Warré hive, you will get a little less honey from them than if they were living in a conventional Langstroth hive.

There are some specific reasons for this, but the overarching one is that the two hives fall on different sides of Mr. Micawber's famous balance sheet of happiness:

> "Annual income twenty pounds, annual expenditure nineteen six, result happiness. Annual income twenty pounds, annual expenditure twenty pound ought and six, result misery."
>
> CHARLES DICKENS, *DAVID COPPERFIELD*

The Langstroth beekeeper's ease of access to those framed drawerfuls of honey allows them to be removed individually whenever the beekeeper feels like it. This design feature allows the level of stored honey in the Langstroth to be manually reduced so that burgled bees are "encouraged" to store more. If honey is the rent the bees pay the beekeeper for their accommodation, Langstroth bees are always slightly in arrears. And like us, if we're already working flat out, the only way to make the rent if we're behind is to economize elsewhere. For us it might be forgoing holidays, restaurants, the little luxuries, or turning down the heat and donning, even darning an old woolen jumper. For the bees it will be less obvious—they have no frivolity to pare down—they will just have less energy available for everything apart from making honey. And while they may not articulate this as Micawber's "misery," they will understand his accounting.

The philosophy of the Warré hive is to allow its bees to store surplus honey *before* the beekeeper then harvests that surplus. This allows its bees to pay the rent to the beekeeper from savings, not with a credit card whose repayments they're constantly struggling to meet. Their surplus honey is an indication that they have enough resources to meet all their other needs in ways *the bees* consider best, not the beekeeper who's farming them from the perspective of human profit. Although the Warré bees might have difficulty with Micawber's concept of "happiness," or indeed anything philosophical, they will experience a quality of life that affords them greater resilience: they'll have a better chance of being here next year.

And for those of us who do sometimes ponder the extent to which we're happy, continuing to be alive is often considered pretty foundational.

Another more rigorous form of accounting was being developed around the same time as Mr. Micawber first proclaimed his economic model, just when Langstroth patented his hive, and Charles Dickens's illustrator George Cruikshank etched his representation of British society as the British Beehive. In 1866 Victoria, the queen at the top of the British social order, knighted William Thompson, an academic working in Glasgow who would become Lord Kelvin. Along with other leading scientists, Kelvin had spent decades working on the laws of thermodynamics. The balance sheets of energy were being defined.

The first law states that energy cannot be created or destroyed—it can only be transformed from one form to another. Plants transform heat and light from the sun into the chemical energy of the sugars in nectar, which bees can consume and biochemically transform into wax. Wax production transforms considerable amounts of energy: one ounce of waxen honeycomb needs the energy provided by about eight ounces of honey. And each ounce of honey requires enormous amounts of kinetic energy to power the flight of its gathering. Because the Langstroth hive is designed to maximize honey production, it follows that it's also designed to minimize wax production. The way it does this is by allowing honeycomb to be recycled by the beekeeper and reused by the bees.

Those frames that surround the combs filled with honey don't just make them easy to remove from the Langstroth hive; they're also reinforced with thin metal wires that allow the framed wax comb to survive being slotted into a centrifuge and spun at high speed to remove the honey like water is removed from our wet clothes in the spin cycle of a washing machine.

The beekeeper can then replace the fully formed but now empty honeycomb back in the hive, ready for the bees to refill with honey. Saving them all that energy they would have had to transform from nectar into new wax.

When you harvest your Warré hive, its free-form combs hanging from their top bars can't be so conveniently centrifuged: no frames or metal wires means that though the honey can be extracted, the honeycomb can't easily survive. The hexagonal cells would be damaged, and the comb couldn't be returned to the hive for reuse. Like trying to remove the cream from a pavlova using your spin dryer and hoping for intact meringue.

So the first-draft accounts for the Warré would show a much greater loss of energy for the hive, as compared to the Langstroth, on wax expenditure.

But if we audit a little more carefully, those first-draft energy accounts don't allow for depreciation: comb doesn't last forever and can't be reused indefinitely. In the Langstroth hive, the nursery is perpetually contained in its lower box by the queen excluder. And here the wax efficiency comes from constantly reusing cells of the comb to accommodate babies: when a youngster hatches, its vacated cell will be scraped clean and varnished with a coating of propolis. This sticky, dark brown, phenolic-scented gum is made from resins the bees collect from the surfaces of plants' leaves, stems, and buds. Plants cover themselves with an almost invisible layer of resins that forms a protective barrier against the bacteria, fungi, viruses, and yeasts that would attack them, and the bees exploit this natural, vegetarian disinfectant, adding a few ingredients of their own to boost its efficacy. Inside the hive it is warm, humid, and sugary—a perfect environment for the invaders those plant resins were specifically exuded to repel—and the bees coat every speck of every surface in this sepia shield. Added propolis is what turns snow-white, brand-new comb into the familiar golden color of the beeswax in the candles we burn, and accounts for much of their beautiful scent.

Which must also help disperse the odor in a cell that hasn't been cleaned by an adolescent holed up in there for two-thirds of his childhood. Once this cell has been thoroughly propolized, it may be repurposed as a pollen store, or a nectar vat, but sooner rather than later in the Langstroth brood box, it will be home to a freshly laid egg, which will develop into another adolescent, who will eventually hatch—leaving the place no cleaner than did its predecessor.

I have known some male adolescents who seem to believe that simply turning their underpants inside out every week or two equates to "dry cleaning." But what might start out as a matter of opinion on standards of hygiene can, after enough cycles of inversion, become toxic to the point of harm.

With every cycle of propolized cleansing, the comb cells get darker—sepia seeps into cesspool, and the shortcomings of the propolis can be exposed: modern pesticides accumulate in the comb, chemicals some beekeepers use against hive parasites linger, and sometimes there is just a dangerous buildup of traditional filth—which of us has ever cleaned anything perfectly?

Old, blackened comb needs to be replaced, and Langstroth bee-keepers will replace it, but a honey maximizing imperative that is eager for comb reuse will tend to keep comb in circulation for as long as possible—and in the confines of the exclusive Langstroth nursery, the renewal schedule is determined by the beekeeper and not the resident bees—who may know better when it really is time for those underpants to go in the bin.

In the Warré hive's hamster wheel circulation of its four boxes, the brand-new comb in the bottom box will become the top box of dark comb filled with honey that you remove four years later, if you harvest a box every year. And though this regular comb removal might err on the side of early comb replacement, the energy cost might be offset by the improvement in hive hygiene and consequent colony well-being and survival.

Although Lord Kelvin was working on the laws of thermodynamics in 1866, it wasn't for his achievements in that field that Queen

Victoria knighted him, but as an industrial pioneer: he was not only instrumental to the successful laying of the first underwater trans-Atlantic telegraph cable, he invented Kelvin's mirror galvanometer, which increased the sensitivity of detection of the Morse code messages being sent through the copper wire that spanned the Atlantic and made accurate communication at such a distance possible.

Kelvin's first contribution to the project was to suggest that improving the purity of the copper in the encased wire would improve the data capacity of the cable.

The uninterrupted comb built by wild bees nesting in a tree cavity is the equivalent of copper wire of a purity that would have extreme hi-fi buffs salivating as they used it to connect their loudspeakers' woofers and tweeters. The bees' feet that walk on this comb are the equivalent of Kelvin's mirror galvanometer—they can detect the buzz of the tiniest Morse code–like vibrations that deliver messages transmitted by other bees gripping onto the comb and vibrating their bodies—at 230–270 Hz the resonant frequency of the wax comb is tuned to the bees' vibration sensors and performs like an information superhighway. It's impossible for two workers to collaborate without communication, let alone sixty thousand—in human commerce the collaboration between so many individuals is now unthinkable without the Internet.

Both the Warré and Langstroth hives interrupt the comb. It's the only way we can harvest honey without killing bees. But minimizing the interruption, as Kelvin minimized impurities in the copper of his wire, will maximize the capacity of the wax comb to resonate and increase the accuracy and extent of bees' ability to communicate.

The wooden top bars in the Warré may interrupt the downward flow of the combs, but they are directly amalgamated into the comb structure, which is fixed to the sides of the hive boxes as it would be in a tree. The wooden top bars will resonate differently than the wax that is custom-built by the bees to their own frequencies, but the top bars *will* resonate and, if left undisturbed, can become an integral part of the acoustic landscape.

For the frames in the Langstroth hive to be easily removable like drawers, they have to not stick. Which means that, however small, there tends to be an air gap between the frame and the walls of the hive. Anyone who has lived beside a noisy road and installed secondary glazing to reduce the din coming through the windows will know that the bigger the air gap between the two panes of glass the better—sound does not resonate well through air. So each frame of comb is more acoustically isolated from all the others in the Langstroth.

Imagine how a violinist might react if you suggested carefully taking apart the body of their Stradivarius and instead of permanently fixing it all back together as before, just resting the various parts next to each other and lightly holding them in place—so the violinist could still play it, but it would be easier for you to dismantle and take a look inside whenever you felt like it. Sound good?

And if whales can communicate over hundreds of miles by vibrating the relatively dense medium of water, I'd hazard a guess that in both the tree cavity and the Warré hive, communication takes place horizontally, between neighboring combs through the resonant medium of the wood their sides are directly attached to in both cases. Otherwise hive-wide communication would be dependent on bees jumping from one comb to another and repeating the message—with all the inherent risks of the Telephone Game—not a great evolutionary strategy. Unlike the whales' oceanic scale of communication, messages sent by the bees through the tree or the Warré hive walls would only have to travel a quarter inch—that space needed for two bees working back to back—before making it to the next wax comb. Coincidentally this is the same distance they have to overcome to pass through the wood of the Warré top bar, so maybe this is a specific level of wooden interruption they understand, whose interference they can factor in and "tune out."

Kelvin's mirror galvanometer was so sensitive that its operators not only could decipher the Morse code signals made up of long and short pulses, they could also detect flaws in the undersea cable that caused interference to the signal. Once recognized, allowances could be made for the effects of consistent flaws—they could be tuned out and the message clarified. But even the best operator of the finest mirror galvanometer couldn't deal with a gap in the cable—communication breakdown. And though bees are probably superior decryptors of weaker signals, even they will be flummoxed by air gaps in their comb created by a Langstroth frame.

Where harvested comb connected to its idle Warré hive wall

So maybe you've heard enough already—you've been convinced that a slightly lower projected annual income of honey is worth the long-term investment it allows you to make in the well-being of your bees—the planet even—for very little effort. You never really planned to become a honey baron. You've committed. You've purchased a Warré hive, or even, with the skillset of someone who can just-about-sort-of-put-up-a-shelf, made one with surprising ease from plans freely available on the Internet as detailed in appendix 1.

And there it stands before you—a stack of empty boxes still trying to pretend they're an injured tree. So what about the bees? How do we turn these boxes into a hive of activity?

Defer your gratification a little longer. Before going any further, we need to put the idle into the idle Warré hive. And we aren't done with Kelvin yet. He gave us the laws of thermodynamics: we've only looked at one. Let's go plural.

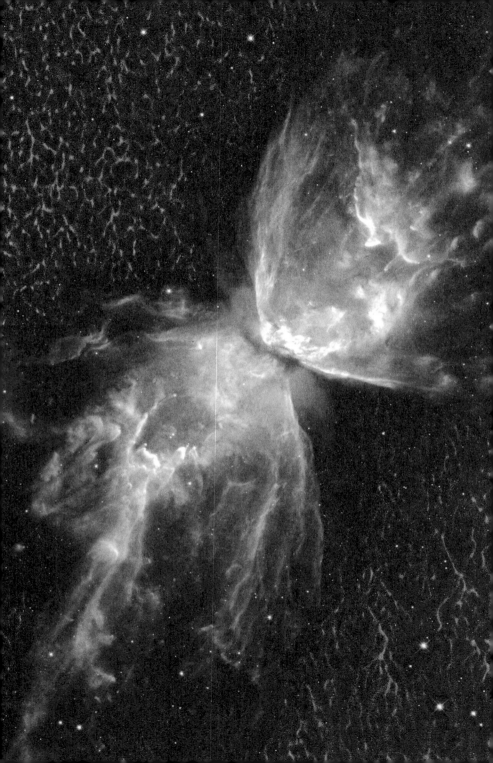

4

A Warm Welcome

Entropy is often simply described as the tendency for things to fall apart. Our lives are full of it. We expend huge amounts of energy trying to stem the disorder that is the harbinger of chaos. We make lists, plans, schedules, laws—even laws about entropy. We tell ourselves stories, forge philosophies, practice religions, and when none of it works, we blame ourselves and resolve to try harder: expend more energy, be more organized, sharpen our focus. Like good citizens of all the great human civilizations that have declined and fallen. Entropy plays a long game.

But maybe we should be a little kinder on ourselves and each other—as Kelvin and his colleagues discovered in the second law of thermodynamics, our battle against entropy isn't just personal—it's a struggle with the universe. And though of course it's never about the size of the dog in the fight but the size of the fight in the dog, the universe is quite an opponent to pick. Even for bees.

On this intergalactic scale, trying to comprehend the logic of entropy really brings out the oxymoron in me, but here's my wet-brained attempt: we think we understand beginnings, at least the need for them, so in the beginning there was a Big Bang; whatever mattered was unimaginably hot and flew apart from itself in every dimension at mind-warping speed; since then everything has been cooling down—into stars, planets, us—and getting even more far flung. We don't really understand why this high-speed fling is actually accelerating instead of slowing down or even maintaining a

OPPOSITE *Hot star*

steady speed. Maybe dark matter will enlighten us. But anyone who's ever drunk a cup of hot coffee can grasp the concept of things cooling down over time. Without having to resort to cloaked phantasms of theoretical physics and without feeling they're plummeting headlong into the depths of outer space—even when the caffeine kicks in.

What physics does tell us about this cooling process is that it is not just something that happens often—to many or even most cups of coffee—it obeys a law that applies to all of them always.

If we were to take our cooling cup of coffee into Mr. Micawber's Victorian parlor, ram a cast iron poker into his blazing coal fire until its tip was red hot, and then plunge it into our cold coffee, would the coffee freeze? And the poker start to glow white hot? Not in our universe! The coffee would be warmed up with the heat transferred from the poker, which would cool down. In our expanding, cooling universe, entropy dictates that unless you're adding energy, heat will always flow from the hotter to the cooler. Not the other way around.

Professor Stephen Hawking was an advocate of the Big Bounce theory in which the universe cycles between expanding and then contracting back in on itself for another Big Bang. In a contracting phase of this universe, everything might go the other way to what we're used to: time might go backward and we might need fire to freeze our iced coffees, but there is little likelihood of this happening soon. Certainly not before you get your bees.

So why should any of this matter to them?

You are rightly hoping that your empty Warré hive will soon become a home-sweet-home to some happy bees. But a better description of the accommodation they need might be "neonatal ward." The queen bee of your colony will be itching to start having babies the second she moves in. Comb building to accommodate this will be frantic, and once the colony's up to speed those daily two thousand eggs are the equivalent of your queen's own body weight: in human terms, a hundred and twenty pound woman giving birth to a healthy eight-pound baby would need to deliver fourteen every day to match. The very thought of this leaves us speechless: we don't even have a word for it—as a species we've currently peaked with octuplets for one day only.

But the queen bee will keep on laying her body weight tomorrow, and the next day . . . building up a brood of twenty days' worth of two thousand eggs per day before the first ones hatch as adults. Supported by twenty thousand adult bees, the hive neonatal ward needs to accommodate up to forty thousand youngsters at the height of summer, with new arrivals replacing those maturing as adults every day. These developing larvae need to be nurtured like human premature babies in hospital incubators—they, too, have a lot of growing to do before they're able to survive in the outside world. And in the same way that human premature babies aren't ready to regulate their own body temperature and rely on their incubators being kept at a steady 98 degrees Fahrenheit, baby bees must be permanently kept at a steady 95 degrees for every day of their development or they'll die—all forty thousand of them.

With only three degrees of separation, your hive is more of a womb than a room.

The nurse bees tending to the larvae can't simply plug in the electric heating pad in a neonatal incubator to keep their babies warm; they have to rely on heater bees to press their bodies against capped comb or climb into a cell in the comb left empty especially for them. Surrounded by cells filled with developing young, the heater bees flex their enormously powerful wing muscles—ones that lift their wings up, and their equal that pull them down. By engaging both sets of muscles at the same time, their wings might shiver a little, but the bee goes nowhere. Instead of the bee's intense effort powering flight, its wing muscles generate heat. When Usain Bolt broke the one hundred meters world record at the Berlin 2009 World Championships to become the fastest man alive, his energy output was 14 watts for every pound of his body weight over those remarkable 9.58 seconds. Every flying bee can generate 250 watts per pound of their body weight over routine six-mile round trips to collect food: if they stay at home, they can donate all that energy as warmth.

Like us, the heater bees don't need to understand the second law of thermodynamics to deliver heat to their babies—all they have to do is snuggle up to them to ensure that life-giving heat will flow from the hotter heater bee to the surrounding cooler larvae. What the heater bees do need is a replenishing source of energy that their muscles can biochemically transform into heat and that is delivered to them by a retinue of feeder bees in the form of honey—transformed sunlight.

But there are times when that solar energy can be too much of a good thing: at the height of the summer, rays of sunlight beating down on the outside of the hive can make the inside too hot for the babies. The heater bees stand down to allow the larvae to be cooled by other bees who fly out of the hive to collect water that is used to douse the comb where the babies are getting uncomfortably warm. This drenched nursery is then fanned by a flurry of bees' wings in order to make the water evaporate. In evaporating, the water moves from a relatively ordered molecular state, a liquid, to a more disordered molecular state, a gas. As a result of this evaporation, the comb the water was resting on will cool down, leaving the babies cooler and dryer.

The refrigerators in our kitchens work in much the same way. But if the inner workings of your fridge are as alien and mystifying as the laws of physics, you can just lick the back of your hand and gently blow on your saliva to feel the cooling principle.

To empathize with the bees' off-grid aerobic aircon let's imagine we're camping in a tent with a newborn baby. Our tent is pitched in a lovely English meadow, near a babbling brook, under a clear sky, on the day in June with the most sunshine—the summer solstice. Proud of the choices we've made so far, we tighten the last guy rope at around ten a.m. and install our baby into the tent only to discover that, somehow, we've left all the bedding at home! We've brought the imaginary food and the imaginary nappies, but nothing to wrap the little darling in—it's as naked as a bee larvae—and yet our baby is happy, not too hot, not too cold. A Goldilocks moment.

But rapidly we discover it is just that—a moment—because the sun shining so beautifully outside the tent has started to warm it up on the inside and our baby is getting too hot. Fanning with a muslin square works for a little while, but the sun is getting more powerful and eventually our overheated baby starts to cry.

We rush to the nearby babbling brook to get water to cool the baby down, panic as we realize we've brought nothing to carry the water back in, and then whoop with delight as we remember—we're wearing imaginary welly boots! Which we take off, fill up, and slosh barefoot back to the tent. Where we pour welly water over the baby, fan it furiously with as much muslin as we can muster, and mercifully, as the water starts to evaporate, the crying stops.

Provided we keep going. As we get closer to two p.m. the trips to the babbling brook become sprints there and back—when we're fetching water we're not fanning—and any reduction in the fanning causes the baby's temperature to rise inside the tent, which the sun has now turned into a sauna. Keeping the fanning going now feels like a hyperactive perversion of Bikram hot yoga—we're so parched we're drinking straight from the welly—but it's keeping the baby alive, and we seem to be getting over the worst of it as the sun heads toward the western horizon.

By around six p.m., so little fanning is required that we can feed our now dry and comfortable baby. Exhausted, we change its diaper as it drifts off to sleep, and put our wellies back on: "How can water be so HEAVY!" Too tired to consider cooking, we wolf down a cold supper.

But as the sun sets, it becomes increasingly chilly. Now we light our imaginary camping stove in the deluded belief that it might warm the tent. The glow of its ineffective flames brightens as darkness begins to fall, the baby wakes up, shivers, and starts to cry again. We don't have time to feel the cold—we're too busy doing burpees and press-ups till we're red-faced and breathless, then taking turns to wrap ourselves around the baby to keep it warm. Making room for our hot and sweaty partner to envelop the newborn when we start to shiver, and numbly dragging ourselves back to the aerobics, like bellows restoring the roar of Mr. Micawber's poker-heating fire.

For over eight hours. Until, sometime after it's risen, heat from the morning sun slowly allows another brief, golden, temperate Goldilocks moment in a summer of *Groundhog Days*.

The second law of thermodynamics universally applies to both bees and fantasy campers: when it's hotter outside the tent or hive, heat will flow into the cooler inside. When it's cooler outside, heat will flow out of both tent and hive. There's nothing we can do to stop it.

Even the legendary King Cnut who set his throne on the seashore and commanded the tide to stop flowing toward him only performed this stunt to demonstrate to his courtiers that royalty had no power over eternal laws—he wanted to show them that the king's feet would get as wet as the next man's if he tried to flout the inevitable.

But Cnut clearly didn't have watertight wellies. If he had, he might have tried a little more negotiation because sometimes a more nuanced approach can find a little wriggle room in those eternal laws: for heat, we can delay its relentless progress from the hotter to the cooler—we can slow the flow with insulation.

A great example of this is the thermos flask, which keeps our coffee hot by insulating against the three ways that heat can leave it: convection, conduction, and radiation. If we pour half a pot of hot coffee into a cup, the other half into the insulating flask, and come back in an hour, the coffee in the cup is cold but in the flask it's still piping hot. Come back in twelve hours: all coffee cold. The insulating effect of the thermos can't stop the heat from the hot coffee from flowing out into the cooler surrounding air, but it slows it down. It resists the flow.

And it doesn't care which way that flow goes: if you substitute ice-cold martini for hot coffee, the thermos will resist the flow of heat from the surrounding air flowing into your deliciously chilled cocktail and making it undrinkably warm.

If keeping their babies at a constant temperature is so important to our bees, should we be housing them in a giant thermos? Probably not: before glass was blown and silvered, bees lived in more naturally occurring homes, and we're really trying to accommodate their successful, long-established preferences. So the inside of the wooden walls of your (still) empty Warré hive are an attractively familiar interface for the bees—so far, so good.

But let's take another, closer look at those tree cavities wild nesting bees still choose as homes. In chapter 2 we saw that the "walls" of a cavity in a tree have to be strong enough to still support the weight of the tree above them—especially when swaying in the wind—and that typically this requires an average wall thickness of around six inches, but the thicker the stronger, as in the cavity with twenty-nine-inch-thick walls.

Check out the thickness of your Warré hive boxes—if you're lucky, they will be a whole inch. And though your Warré hive doesn't need to carry the weight of a tree, it does need to insulate its temperature-sensitive occupants.

There are various ways we can measure the thermal resistance insulating materials exert against the flow of heat, but one of the simplest is their R-value. "R" conveniently stands for *resistance,* so the higher the R-value of a material, the higher its resistance to the flow of heat and the better it insulates.

Your Warré hive box's one-inch-thick walls have an R-value of R1. And yes, that is low. Alarmingly, R1 is about the same R-value as a canvas tent. Like the one in which we just imagined that twenty-four hours of fun with a newborn.

R-value is cumulative. So if the walls of your Warré hive were a strong ten inches thick instead of one, they would have an R-value of R10, which would make them much better insulators. And R10 is a common level of insulation that bees living in average tree cavities have successfully assumed they could rely on to help them keep their babies not too warm and not too cold for millions of years.

So should we add an additional nine inches to make the wooden walls of our Warré hive boxes ten inches thick? Probably not: we'd be increasing the weight of the hive box from less than the weight of a newborn baby to more than that of an eight-year-old child, or a full five gallon watercooler bottle—forty pounds is heavy when empty, but when that box is full of honey it might weigh as much as forty pounds more. Harvesting eighty pounds—carefully lifting the weight of an eleven-year-old or two full watercooler bottles off your hive—risks marring a delicate procedure with spine-tingling clumsiness. Let's file it away under "Osteopath's Option." Could there be a Goldilocks insulation solution? Not too unnatural, not too heavy, but that gets that R-value of R10 just right?

Although it has only been around for eight million years—a whippersnapper compared to the bees—there is another animal with super powers that can help us: this creature is capable of climbing vertiginous mountains and miraculously transforming the tiniest amounts of barely edible scrub into temperature-specific insulation: wool.

Sheep grow it to spare, and we can harvest the surplus to insulate our hives. Because it has an R-value of R3.5 per inch, wrapping our hives two and a half inches thick in wool will raise their walls to R10 while adding next to nothing to their weight.

Of course, there are other materials that are very effective insulators. If we were to hold a casting session for the part of "sophisticated, hardworking, tree-related hive insulator" in our coming-of-age bee drama, we might consider a few actors: wood and polystyrene as well as sheep's wool. When asked how they see the part—what's their primary motivation—they might give us very different responses:

Wood: "I'm here for the tree—whatever support it needs, I'm there."

"How about bees?"

Wood: "Bees? No, I do *leaves*. It's all about getting the leaves up to the sun."

"For warmth . . . ?"

Wood: "That's up to them—I just do the height."

Polystyrene: "Well, I've got relatives who used to be trees, but now I mostly mix with other chemicals—it's good—they expand my horizons."

"And what's that expansion good for?"

Polystyrene: "Er . . . I dunno really . . ."

"Well, what can you do now that you couldn't do before you expanded?"

Polystyrene: ". . . Squeak?"

Wool: "I know nothing about trees! I stick to sheep—they need my help, so . . ."

"Help with what?"

Wool: "Staying alive."

"Sounds dramatic!"

Wool: "It can be—they pretty much have to stay at the same temperature whatever the weather and I can help with that."

"That's hardly life or death."

Wool: "Well, you'll know this for yourself—if your temperature gets, what, 7 degrees Fahrenheit above normal, how do you feel? Awful! You can't function. Same if it goes 7 degrees below—you can't think straight, everything stops working. And if those 7 degrees get to 14 you haven't got fever or hypothermia anymore—you're dead. And if my sheep's dead, I'm out of a job."

"So how is it you help?"

Wool: "I trap air."

"That's it?"

Wool: "I tailor the air-trapping to the temperature my sheep likes best—happy sheep, happy me."

"Tailor?"

Wool: "Sure—I've got springy, spirally, hygroscopic fibers that expand and contract when the humidity changes, and they allow me to . . . Anyway, enough about the sheep—I'm really keen to branch out—tell me more about the part . . ."

The Warré hive becomes an idle Warré hive when we wrap it in wool. If our imaginary tent had been wrapped in two and a half inches, increasing its R-value from R1 to R10, that wool would have allowed us to be more idle—it would have slowed the flow of heat both into and out of the tent. So less heat energy would have flowed in during the daytime before evening fell: the time when the direction of flow changed. Then less heat energy would have flowed out during the night before the sun rose to continue the cycle. This would have curbed both extremes of temperature inside and reduced the insane amount of frantic fanning and bellowing burpees we had to do to keep our newborn comfortable.

In determining the inches of thickness of wool, 2.5 is not a number chosen by a conspiracy of Illuminati numerologists, nor do I secretly work for any Wool Marketing Board. None of my immediate family are sheep. The wool cannot "heat up" the bees, let alone "cook them"—it brings no new energy into the hive on a cold night, nor does it remove any on a hot day. It is merely a lightweight, low-carbon, renewable, biodegradable attempt to give the bees the level of insulation they've grown accustomed to through the course of their evolution while they successfully regulate their *own* temperature with awe-inspiring accuracy.

You can of course just wrap your hive in old blankets, jumpers, scarves, and even socks—the bees won't care. But because over the course of the year the hive's height will vary, it's better to insulate each hive box individually as an integral part of its construction. On the outside of all hive boxes, beekeepers add handles so they can conveniently lift them. The idle Warré hive adds something for the bees' convenience: a sleeve of wool covered by overlapping breathable roofing membrane to keep it dry. Any bee-respecting hive box should feel naked without it.

Appendix 1 gives instructions on where to buy and how to make a Warré hive, and appendix 2 describes how to insulate your idle Warré hive. Your bees aren't expecting haute couture, so until imminent off-the-peg solutions become available, well-meaning, cozy approx-

imations with a style all their own are more than good enough. And you only have to apply it once: this insulation is permanent. Some beekeepers are nervous about leaving it on in the summer in case the bees "get too hot." I promise them I will remove my hives' insulation over the summer the moment they show me a tree that spontaneously reduces the diameter of its trunk by eighteen inches every spring and regrows that eighteen inches every autumn. And keeps doing it even years after death.

I'm not even so persnickety as to insist it has to have bees living inside it.

Insulating like this will also increase your chances of getting free bees, but if it's certainty you require, you'll have to buy some. I've done this, and though it's not my preferred way to get bees, it delivers: you can order bees online in December and they'll arrive in a neat, ventilated, bee-tight box at your front door the following June. It's best to find a supplier who's as geographically close to you as possible so the bees will experience the minimum change of environment from the move and settle into their new home with you more quickly.

What you will get inside the humming box the postman wants your very quick signature for is a "nucleus" of bees. Five frames of populated honeycomb—a fully functioning microcosm of a colony—not just ready to go, but already a work in progress. A queen will be laying eggs and about five thousand bees taking care of babies at all stages developing in comb that's filled with working stores of pollen, nectar, and even honey. This nucleus just needs to be inserted into your hive and away you go, beekeeper.[1]

But you might have tripped up on the word *frame,* and indeed that is the rub with this route to bees for the idle Warré hive: currently almost every available nucleus of bees is tailored to beekeepers with conventional Langstroth hives with frames. So they can slot the new bees in quick as shutting a drawer. The problem we idle beekeepers have is not that these bees won't be delighted to expand downward and build more free-form comb in our frameless hives, it's that the frames they arrive on simply won't fit into the idle Warré hive. They're too long.

But it's simple to make a converter box that will accommodate your five frames and sit atop your idle Warré hive. The converter box is partially open at the bottom, allowing the bees to naturally build new comb downward and thereby inhabit your idle Warré hive at their own speed. Appendix 3 shows how to make one.

The bees in the nucleus you're left holding by the retreating postman have been cooped up for longer than they'd like, but before transferring their frames of comb into your converter box, it's a good idea to ease their arrival by moving your idle Warré hive aside from the location you've carefully chosen and positioning the nucleus box so its heavily taped-up entrance is in *exactly* the position your hive's entrance will occupy. Go away and have an idle cup of tea for at least

---

1   It's possible to buy a "package" of bees—a box of a random assortment of workers with a random queen to whom they have no relationship—she's caged to stop the workers from killing her—and no comb or frames. Packages are expensive and offer the worst chance of a viable colony. Not recommended.

half an hour or so to let the bees calm down—the tree they think they're living in has been behaving with insane unpredictability for the past day or two, and even though they're keen to get out and get on with flying to fetch food, a little enforced, reorienting period of stability is no bad thing.

You're now about to have your first free-range contact with your bees. You're going to open the entrance of the nucleus box and give them access to your world. This includes you. So you might want to wear a bee suit and some gloves. If they help you feel confident, that's a good thing. However tough or brave or calm you want to be, if you are nervous you will involuntarily be giving off stress phero-mones—you can't help it and you can't smell them, but the bees will, and they might think you're stressed because you're a threat, and if they think you're a threat they might misinterpret your behavior and sting you. If even thinking about this makes you stressed, wear the suit—it doesn't have to be forever—just until you know your bees, and then you can dress or undress as you like. Even if you're an expe-rienced beekeeper, you don't yet know these new bees and, just as important, they don't know you.

I've never washed my bee suit. Beekeeping can be sweaty work, and my kids tell me my bee suit is rank. I keep it that way because it lets my bees know, in no uncertain terms, that the person wearing it is me, and I don't hurt them—like it or not, my odor has no historical association with harm, it doesn't smell like trouble's coming. But to new bees that don't know me the suit is a wise, if stinky, precaution: besides the barrier of the fabric, it offers a powerful pungent disguise to any stress pheromones I may inadvertently release, reducing my detectable fear on the Beaufort scale to "a fart in a gale of wind."

Suitably clad, carefully peel off the tape, open the entrance, and slowly retire, allowing your bees to fly freely and familiarize them-selves with their new surroundings. Within an hour you'll probably see bees returning to the nucleus box with little colorful blobs of pollen on their hind legs: they've found food, they've got shelter, and you can leave them like this for a few days—they've been thriving in this nucleus box for weeks—whilst you hope for ideal weather conditions to move them into your hive.

If you find that your new arrivals are a little cranky after their travels, and that the guard bees are being more officious than you'd like when you approach the nucleus box, make a simple effigy of yourself from some unwashed trousers and a ripe shirt that you can hang in the breeze ten feet from the hive entrance: inquisitive guard bees will investigate it thoroughly for maybe a day or two, and then lose interest as your smell becomes part of their landscape. Your diplomatic pheromonal stand-in can then retire to the laundry as you nonchalantly approach to rehouse your bees.

A still, dry, warm twilight is good because most of the foraging bees have flown back into the nucleus and the colony has gathered together for the night. Before disturbing the bees for the transfer, some beekeepers like to give them a little smoke to "calm them down," but I prefer to wait and see what their mood is and, if they're agitated, gently mist them with water from a hand sprayer: it's kinder to the bees and at twilight they're usually pretty docile.

Move the nucleus box to one side and restore your idle Warré hive to its chosen position so its entrance is *exactly* where the nucleus's entrance has been. Open its lid and slowly, carefully lift the frames out of the nucleus box one by one and place them into your converter box. Keep them in exactly the same order and exactly the same way around—they are functionally connected to each other by the bees in ways we only partially understand: imagine what your electrician and plumber, let alone your plasterer and flooring contractor, would have to say about your thinking that you can simply swap your kitchen and living room. Once all the frames are correctly transferred, softly cover them with a piece of sturdy fabric cut to the size of the converter box to gently persuade any bees wandering around on top of the frames to go down between them. This fabric will not only tuck these bees in now, it will also help with the harvest to come. Then rest the roof in place on top slowly enough for any remaining bees to be nudged out of its way and avoid being accidentally squished.

There will be a few tenacious bees that are reluctant to leave the nucleus box—it still smells of their queen and to them that means it's still home. Place the opened nucleus box near the entrance to your idle Warré hive, and eventually the queen's scent emanating from the hive she's now inhabiting will overpower her fading traces in the nucleus box, and the last of its bees will join her.

You can watch my first beginner's encounter with a nucleus at www.youtube.com/watch?v=nAzL_VPGgNs.

Leave the converter box on top of your hive until the following autumn when, hopefully, like any box on the top of an idle Warré hive, it will be full of delicious honey that you can harvest by removing it, leaving your bees hard at work in the boxes below. With the hive now insulated with wool on the outside and finally frameless on the inside, you will be a fully fledged idle Warréor.

5

Bee Chosen

If ordering bees in December and having to wait six months before they arrive with an invoice feels beyond the limits of your patience, you can always steal a march and try to catch a swarm for free. Traditionally,

"A swarm in May is worth a load of hay,
A swarm in June is worth a silver spoon,
But a swarm in July is not worth a fly."

The rhyme's valuations are based on the swarm's chances of surviving the coming winter, and the bees in that swarm will have made the same calculation before they decided to leave the hive where they came from.

Swarming is risky and not compulsory, but it is the way one colony becomes two, and the drive for increase is primal and urgent in all living things. In the spring a bee colony will be monitoring itself and its environment to assess if both are healthy and robust enough to give favorable odds for it to divide itself in two, giving both parts a good chance of survival. The earlier in the year the colony has been able to build up its numbers to a critical mass where swarming becomes feasible, the longer both the departing swarm and the bees left behind have to collect enough food to store as honey to see them through the winter.

OPPOSITE *This one—quick!*

If the colony divides when there are too few bees, no matter how much food is out there, in both of the two new colonies there won't be enough bees to collect enough food to enable their queen to turn it into more food-collecting bees—the exponential growth of population needed to make the numbers of the survival equation work.

If the colony waits too long before dividing, there may now be plenty of bees to collect food for each of the two new colonies, but enough food may not be out there anymore—too many plants may have already gone to seed.

So when a colony does commit to divide, they don't dither: a swarm of as many as twenty thousand bees flies out of their hive and clusters protectively around their queen, hanging from a branch, a pub sign, traffic lights, a stationary bicycle, or anything conveniently nearby while scout bees are flying off to find a suitable new home. And when the bees decide where that new home will be, you want their chosen des res to be one you've offered up: a bait hive.

Bait is a word that even the most voracious real estate agent would be reluctant to use when advising us how best to sell our homes, but putting ourselves in the position of house-hunting bees might help us understand which "tree cavity–like inducements" might help to persuade them to move in. And we can draw on some peer-reviewed market research into their preferences.

Bees seem to like a home about fifteen feet above ground level. This height will discourage ground-based predators from disturbing them, but don't let it discourage you: if fifteen feet is above your reach, pick a site for your bait hive as high as is safe—this doesn't have to be where you plan to keep your bees permanently. Any higher than fifteen feet in the wild and there's an elevating risk of living inside a thinner part of the tree that might snap off and crash to the ground. They also like their entrance to face south, not only to get immediate early morning navigational guidance from the sun but also to reduce the increased dampness and humidity of a north-facing entrance—being so high above moist ground also helps keep the air at their threshold drier.

Wind- and water-tight with secure access is essential, but bees won't be expecting a set of keys to an insurer-approved lock. Or even a door. They need their entrance to be open at all times for controlled ventilation and access, so for them size is everything: too big and they can't defend it from intruders; too small and the air traffic bottleneck will throttle peak-period food deliveries. An opening you can just slide the long side of a deck of cards through seems to do the trick and entice scout bees to take a look inside.

The scouts are looking for a space to accommodate their family when it grows big enough to be able to store enough food to survive the winter. They calculate the volume of a cavity by pacing it out in three dimensions. Too big a space is not good—imagine the heating bills if you moved your family into a cathedral for the winter—but a volume of about ten gallons seems to be snug. Conveniently, this happens to equate to two Warré hive boxes, so you could make two bait hives out of one standard Warré hive to double your chances of catching a swarm. For now, leave the top bars out of the lower of the two boxes—if you leave them in and the scouts don't happen to crawl through their gaps, the scouts might be confused into thinking these lower top bars are the ceiling of a cavity half the size of the one you're offering, and therefore half as desirable.

Once the scouts have surveyed the uninterrupted inner dimensions of your two Warré boxes and found them to be ideal, the next thing they look for is evidence that this isn't just space of a good size, it's accommodation—it's been successfully occupied before.

Smell is very important. Like human queens who know that everywhere they visit doesn't usually smell of fresh paint but can still appreciate the spirit in which it's been applied, everyone who's heated bread in the oven and made freshly ground coffee before potential buyers visit their newly painted home knows that it's worth the trouble to create an atmosphere of effortless, natural homeliness of the highest order.

For the scouts a cavity that smells of propolis and beeswax smells of home.

An empty hive box that you've recently harvested honey from would be perfect, but in your first year you can buy both—apply drops of sore-throat–relieving tincture of propolis to the insides of your boxes like perfume, then rub on a good coating from a block of the purest beeswax you can find.

Something that not only smells good but will look and feel great too is a piece of old comb. You'll need to get it from a friendly bee-keeper when you're starting out, but think of it like a really fancy designer kitchen—it tells the scouts that successful bees have lived here, and if you can engineer it so this comb is hanging from the top of the space, even badly tied with string, it will be a piece of designer kitchen that's in the kitchen—always more reassuring than expensive kitchen appliances strewn over the living room floor.

Even if you can't lay your hands on some old comb, we can at least set up the kitchen so it's got the equivalent of all the correct electrical sockets and plumbing connections in the right places ready for the most discerning kitchen designer: the Warré top bars on the ceiling of our cavity are already in the right position, but we can make them even more attractive by adding starter strips of wax running exactly down their middle: precisely where the bees will feel most comfortable hanging to start building comb, and where it's most convenient for us.

We can nail strips of wooden molding with a triangular cross-section onto the underside of the top bars. These use gravity to guide the upside-down, comb-building bees toward the point of the triangle, which now makes a sharp, straight edge we've created along the line that's ideal. If, like using a knife to scrape burnt toast, we now scrape the sharp edge at the bottom of the top bar against a block of beeswax, we deposit a tiny amount of wax along its length. It may not seem much, but to a bee it's like a house builder turning up to a greenfield site with some drawings, only to find that someone's dug some foundations and laid a couple of courses of bricks—and weirdly, they all measure up to the plans! There is of course nothing to stop the builder, or your bees, deciding to rotate the plans through

90 degrees and start building in a completely different orientation, but in practice neither tends to be so perverse.

And now we're into active marketing—the equivalent of a real estate agent's "To Let" sign outside the front door can be provided by a few drops of lemongrass oil: this smells the same as a communication pheromone that bees secrete from the Nasonov glands in their abdomens and fan out with their wings to tell passing bees, "Over here! . . . This way!" Drizzle a little lemongrass oil onto a piece of paper tissue you've folded over to the size of a stamp, seal it in a flat package of tightly folded aluminum kitchen foil, and fix beside the entrance to your bait hive with a thumbtack—more than enough of the scent will escape through the hole you've just punctured in the foil, and it will keep trickling out its message for weeks. It's worth spending the tiny bit extra buying organic lemongrass oil—the bees' sense of smell is so sensitive they might be put off by any microscopic amounts of pesticide. Would you choose anything for your family that advertised "Free from nuts, but made in a factory that also handles poison"?

And just in case a passing scout happens to be upwind of your come-hither aroma, visually indicating the entrance so she can see it from afar might be helpful. Ultraviolet (UV) light is the most eye-catching for the bees, and an array of straight lines in UV paint that converge down toward the entrance will turn your bait hive into a beacon. Conformal coating is a clear UV paint used by the electronics industry to waterproof circuit boards, so it won't wash off in the rain and it comes in little bottles with a brush like nail polish. Paint whatever enticing design takes your fancy, but make it bold and clear—only the bees will see it and they're looking for a marker, not a manicure.

So: half a south-facing Warré hive; perched fifteen feet off the ground; triangulated top bars; waxed and propolized; bee-targeted advertising. It may be "bait," but it's not idle: we haven't wrapped it in wool yet, and this could be the determining factor. Because scout bees can measure the level of insulation of a potential future home.

But wouldn't that require a lot of technical equipment and a PhD in thermodynamics? Or at least a calculator, slide rule, or abacus? We pay highly qualified surveyors to produce accurate energy performance certificates that inform prospective purchasers of our houses how well the structures are insulated and how expensive they will be to heat and cool—how could an insect with a brain the size of a grain of salt do that?

Professor Tom Seeley of Cornell University is the author of several books on honey bee behavior, including *Honeybee Democracy* (2010) and *The Wisdom of the Hive* (1995). He was the recipient of the Humboldt Prize in Biology in 2001. He primarily studies swarm intelligence by investigating how bees collectively make decisions. One of the things his research into the behavior of scout bees has revealed is that they make multiple visits to promising cavities, and we want our bait hive to deliver on that promise. Professor Seeley does not claim that scouts measure insulation—that is a hypothesis proposed by a woolly-minded idle beekeeper, but let's entertain it for a moment.

Let's go back to the tent. Yes, *that* tent. But this time it's going to be a bit more relaxed: we don't have our newborn with us—it's being safely looked after by its aunties at home. And although we're back in that lovely imaginary English meadow, near the babbling brook, under another clear sky, on the summer solstice in June—the expanded and renamed "Silver Spoon Glampsite" is now offering us the choice of either *that* family tent or brand-new accommodation right next door that is identical to the tent in every respect, except that while the tent is fabricated from canvas a sixteenth of an inch thick, this new dwelling is made from ten-inch-thick logs—it's a family cabin, and we're here to inspect both before committing to either.

Like scout bees, the only inspection "tool" we can use is multiple visits.

At nine a.m. we make our first visit to compare the imaginary tent and cabin and discover that they look and feel like identical twins on the inside. Even down to little foil-wrapped, high-cocoa imaginary chocolate hearts that have been carefully placed on the plumped pillows. Clearly under new management. The complimentary chocolate looks tempting, but we had a hearty imaginary breakfast and decide to eat it when we come back later.

At three p.m. we walk inside the log cabin: everything looks the same, but it's comfortably cooler than the hot meadow. We sit down on the bed and enjoy those imaginary chocolate hearts—delicious, but rather small . . . so we walk into the tent for more. And walk straight out again, glimpsing the molten imaginary chocolate ooz-ing onto the pillows from the collapsing foil hearts: the tent is like a sauna, unbearably hotter than the hot meadow.

By nine p.m. it's getting slightly chilly on the meadow. We wish we'd brought an imaginary jumper. We walk into the tent, but it's just as chilly as outside. The puddles of chocolate we pick off the pillows are now hard, brittle, and difficult to separate from their imaginary foil with fingers that are becoming increasingly numb. So we walk into the log cabin, which is now significantly warmer than either the tent or the meadow outside: but comfortable, much like it was at nine a.m.

Do we need that PhD in thermodynamics to calculate if it's the tent or the cabin that will be easier to maintain at a constant tem-perature for our baby? By making repeated visits at different times of day, like the scout bees with their brains the size of a grain of salt, we could literally work it out with our eyes shut—we just need to be able to feel the difference in temperature between inside and outside.

But any barrister worth their many grains of salt would be quick to point out an alternative hypothesis to explain the repeated visits by the scouts: they are simply forgetful. They have appalling mem-ories and have to keep going back to remind themselves of what they've just seen. Bees have brains much smaller than those of gold-fish, who can't even remember what happened the last time they swam round the bowl.

The case against goldfish memory is not watertight, and I would submit a collective character reference for the bees.

All honey bees share a very similar career path. When they hatch as adults, they will work inside the hive. For three weeks they will progress through a series of jobs—starting out as cleaners and then moving on to become housekeepers, builders, nurses, heaters, honey-makers, pollen fermenters, comb builders, undertakers, and entrance guards. For the next three weeks their jobs take them out of the hive: they're pilots, navigating the airways foraging for nectar, pollen, and propolis, sharing intelligence about their sources, flying farther from the hive the more experienced they become. Most bees will do most jobs, and it is this direct, personal experience of what every bee needs in order to do every job that qualifies the most senior scout bees to be able to assess the best place for a colony to choose to make its new home.

If scout bees had poor memories, they would make poor decisions, and their species would have been lucky to survive fifteen thousand years. In fact, Professor Seeley has calculated that scout bees pick the best available cavity for a new home with a consistent success rate of over 95 percent, which probably contributes to their still being here millions of years on and confirms that they are well informed and can remember that information. Specifically, scouts were formerly heater bees. And in order to be a heater bee that's helping to regulate the temperature of your babies, you have to possess sensory organs that can detect the temperature of what you're heating—so you can stop when the babies get too hot and start when they're too cold. Babies' lives depend on heater bees' expertise and precision.

And this precision doesn't stop here.

Because the bees have developed the equivalent of a fast-track graduate entry scheme so their workforce can quickly respond to changes in their environment.

Normally bees' career paths are determined by age, which delivers the right numbers of the right professions of bees to deal with normal circumstances. But if, for example, the food supply dwindles because of a locally occurring "gap" in continuing blossom caused by the particular mix of species of flowering plants nearby, the queen will respond by laying fewer eggs, which will require fewer nurses. In the immediate future, what might be more useful than the standard quota of nurses would be more pilots. Pilots could swell the number of foraging bees flying to search and collect food—the more bees you have out there, and the more senior foragers who can navigate farther from the hive, the better your chances of increasing your food supply, even in meager times.

Bees can't control the environment outside the hive, but they can control how their babies develop inside the comb. While they nurture their children into adulthood, the bees can control their development to determine how quickly or slowly they climb the career ladder and how good they are at learning. And this fast-track control is not determined by a special diet as is the case with the development of a queen; it's achieved by tiny changes in the temperature that the developing youngsters are kept at. Like cooking different professions to order.[2]

Heater bees aren't just capable of a single accurate thermostat setting of 95 degrees Fahrenheit—generally healthy for all bees; they can vary their thermostat controls through fractions of degrees to help deliver specifically qualified adults into the workforce as they're most immediately useful.

This flexibility gives bees an evolutionary advantage, but the precision heating requires exquisitely precise temperature sensitivity on the part of the heater bees—who will progress up their own career path to become scout bees. And scout bees who have experience of actively regulating temperature by fractions of degrees will be expertly capable of using the same sensory organs to detect *huge* temperature differences, two orders of magnitude greater, when they fly in and out of insulated and uninsulated bait hives. In the same way an accountant capable of balancing the books of a major corporation can still count on their fingers. Neither accountants nor scout bees are goldfish.

2   M. A. Becher, H. Scharpenberg, and R. F. Moritz, "Pupal Developmental Temperature and Behavioral Specialization of Honeybee Workers (*Apis mellifera* L.)," *Journal of Comparative Physiology A* 195 (2009): 673–679.

Making your bait hive idle by wrapping it in wool will reduce the amount of work it will require from the bees. Every day that the temperature outside the hive deviates from 95 degrees Fahrenheit means more work for bees maintaining that temperature in an uninsulated hive: and around my neck of the woods that means every single day.

Scout bees can sense the lifelong advantages of an insulated hive and will factor the benefits it confers into their choice of a new home. I have no mathematical proof for this and haven't even asked any of them if this is true. I can only offer anecdotal evidence of correlation, which is not causality, but here's my anecdote.

On the front of my house there is a tiny balcony on top of a bay window, directly outside the window of my office. It faces south, and the top of its wall is fifteen feet off the ground. This is where I perch my bait hives. Two of them, six feet apart. They are identical in every respect, except that one is insulated to R10 with wool, and the other is uninsulated bare wood with an R-value of R1.

For three years, when a swarm moved into one of the bait hives, I would remove it and replace it like for like with another empty bait hive. But I would swap over the positions of the insulated with the uninsulated: in case there was something special to the bees about the position of the hives that I didn't know about—an attractive scent I couldn't smell, minor variations in the earth's magnetic field, distance from mobile phone masts, even ley lines!

I contacted Professor Tom Seeley with my results and he was kind enough to inject some professorial rigor into my methods. He didn't like the bare wood of the uninsulated hive—the bees might be choosing the insulated hive because they liked the smell of the wool, or the look or feel or smell of the waterproof roofing membrane that covered it—you never know. So on his advice I controlled for that by pinning wisps of wool thin enough to add as negligible insulation to the bare wooden boxes as a few dabs of Eau de Mouton perfume, and covering them with waterproof roofing membrane so the uninsulated bait hive would now look, smell, and feel identical to the insulated, both outside and in.

Eight swarms in five years is a laughably small sample size, but whatever the relative positions or exteriors of the two different bait hives, all eight chose the insulated 100 percent of the time.

And the choosing is dramatic. It might have been weeks since you saw the very first solitary scout bee arrive at your bait hive, bobbing in a hover of curiosity near its entrance as you willed her to go inside and have a look. She goes in! And then comes out again—too quickly for any commitment? She hovers, then disappears back inside. You crane for glimpses of her pacing the entrance. She flies out and bobs and hovers, examining the side of the hive. She flies off. Now she's back bobbing on the other side. She flies straight toward you— bounces off the glass in your window with a dull click, and flies away.

But she, or a bee who looks very like her, comes back with another scout. And another. And the morning comes when there are more than ten scout bees all bobbing and hovering in and out of your bait hive, and then it's dozens. And the focus and urgency become palpable.

You're drawn outside by something akin to an imminent thunderstorm—a change of energy in the air that you can't place, only feel. A resonance. And then you start to see them, like they've come from everywhere, all ways. One time a man came running around the corner from Portobello Road Market and stood beside me, staring in wonder. He'd been buying vegetables at a stall when he found himself running without really knowing why—he told me he'd felt a strange, positive energy and was drawn to its source. He had no idea that it came from the twenty thousand bees that were filling my street as they orchestrated getting into their new home, he'd just run toward it.

Some people do the opposite. Sometimes waving clothing and screaming as they try to get away from a swarm of bees as fast as they can. Even though the swarm's destination is fifteen feet above street level, sometimes the periphery of the blur of a really big swarm expands down to the pavement. During the swarming season I keep my bee suit and bee smoker close at hand. The combination of the two seems to reassure passersby as effectively as wearing a white coat and stethoscope in a TV hospital drama, but the crucial thing is not to wear the veil—not to disappear behind a protective screen as if you need to be protected. Standing calmly in a white suit, holding an entirely redundant bee smoker, smiling as the odd bee accidentally bounces off you, seems to qualify you to explain to people that although it might look like they're in the middle of a biblical plague of stinging buzzy things—in fact they are probably never less likely to be stung: all the bees flying around are currently homeless—they won't be in half an hour—but right now they have no home, no babies, and no food to defend. Stinging is a defensive, not an offensive, reflex. The only thing a swarm is defending is their queen. And even someone standing flailing their coat in the air to defend themselves from an imagined attack by "killer bees" has got better odds than 20,000 to 1 of not being stung: for the coat-flailer's misplaced fears to be realized, they'd need to hit the queen, and her immediate protective entourage will ensure that her flight path is directly toward the entrance of the bait hive—she won't be allowed near to anyone down at pavement level. Counterintuitively, we are less likely to be stung in the middle of this blizzard of bees than standing too close to an established hive protected by one very determined guard bee.

Once the queen has been ushered into your bait hive, the entire swarm will quickly follow her. In less time than most of us wait at airports' baggage claim, bees that seemed to fill a street will disappear inside the hive, settling into their new home inside half an hour, the last stragglers brushing in past early adopters now nipping out for food.

With the same speed, the beekeeper standing on the pavement is transformed from emergency bee savior to weirdo in white, but though it may be time to swiftly retire from public relations, your work with this swarm is not yet done. Chances are that the place you've perched your bait hive is not the place it's going to stay—what's ideal for the bees may not be the ideal place for you to access and work around the hive, so you're probably going to have to move it to your apiary, or if that title feels too grand, just your carefully chosen bee-friendly place that allows you more convenient access. And you really need to make the move this evening.

Those bees inside your bait hive all reset "Home" on their GPSs as they flew in. The longer you let them stay in the bait hive's original position, the more hardwired its coordinates will become. And like car drivers who follow their GPS's guidance into rivers, some bees can be fatally tunnel-visioned about returning home. If the hive has been moved, even by a yard, they can persist in trying to fly in where the entrance used to be, eventually dying from bewildered exhaustion.

With the setting of the sun your new bees lose their essential navigation reference and will return home. Wait as long after that first sunset as possible before making your move, but don't leave it too late—being able to see what you're doing at the other end of the journey is highly recommended. Even if you built your bait hive to Fort Knox specifications, the psychological reassurance of adding tightly bound strapping is worth considering, and once you're certain the bait hive isn't going to fall apart during transit, it's time to gently contain its twenty thousand occupants and turn them into passengers by blocking their exit. For a short journey a foam sponge tightly stuffed into the entrance will work, but if the bees are going to be stuck inside for longer they'll need more ventilation—cover the entrance in plastic or metal fly screen, carefully held in place with thumbtacks so it's easy to remove on arrival—but if you're driving make sure there really is no way any bees will be able to join you at the steering wheel . . .

My first swarm relocation was quite a learning curve. Or slope. At twilight I zipped up my bee suit in case of disaster, prepared the bait hive for the journey up to my rooftop, and started to climb the stairs to the top floor of my house where there's a small roof hatch only accessible by ladder. I knew the bait hive and I could only get through the hatch together if it was on my head. I was prepared for this. What I'd failed to consider until this moment was that the hands holding a box of twenty thousand bees steady on your head cannot, at the same time, hold on to a ladder as you climb it.

As I was about to put my second foot on the second rung, a bedroom door opened: a familiar voice, a familiar question, "What are you doing, Dad?" My twelve-year-old son was confronted by a vision of a veiled, biohazardous sherpa channeling his inner gyroscope: not a great look, but apparently highly effective at encouraging instant compliance—"Get back in your room!"

The ascent was going smoothly, at least for the bees, until we reached the hatch. I'd squeezed hive boxes through it before and knew they'd fit, but the realization now dawned that I'd always tilted them to align with the slope of the roof. I had no idea if the inhabited bait hive would fit through while remaining vertical. The physical contortions required to attempt this maneuver turned ligaments that were already unwilling victims of free-form Tai Chi into shredded, coruscating kimchi. But eventually the sensation of hatch scraping against knuckle turned into the soothing caress of the breeze of a London skyline.

Once your bait hive is located in your chosen permanent spot, we need to encourage the bees to discover that their home has changed position and hasn't merely been gently swaying in a tree that's caught the breeze. Remove whatever you've used to block the entrance and quickly re-block it with a handful of grass. And not too tight: now you want the bees to be able to escape—eventually. The next morning when they try to leave the hive for work, your new bees will encounter this grassy obstruction. The effort required to move it out of the way should be significant enough for them to know that something new and strange has happened at the hive entrance and, just to be on the safe side, they'd better recalibrate "Home" on their GPSs to their current location. For urbanites and those nervous about the grass trapping the bees inside, two hair claw clips side by side with their teeth filling the hive entrance will create nearly the same level of disorientation as the grass.

Both will cause departing bees to perform orientation flights near the hive entrance to relate its position to nearby landmarks before heading off to gather food. Without the grass or hair clips, more foragers will simply assume home is where it used to be and end up dying when they don't find it there.

The next day, check that the bees have in fact managed to remove the grass from the entrance and are free to come and go—if not, loosen the grass, but still leave them work to do to get rid of it. If you've chosen hair claws, wait until the bees have stopped flying in the evening before removing them.

After about three days, return to the bait hive at twilight with the top bars you left out of its lower box and one additional idle Warré hive box complete with its top bars. Once all the bees are tucked inside for the night, place the additional idle Warré hive box on a stable surface next to the bait hive. Lift off the roof on your bait hive and put it to one side. Check that you can easily separate the top box of the bait hive from its bottom box—the bees may have already started sticking them together on the inside with propolis—if so, use a hive tool, the all-in-one "spanner"  for beehives, to delicately separate them just a hair's breadth—as if you were prying the lid off a can of paint.

Almost all the bees inside will be up in the top box where they will have already started building new comb, and you need to lift this box very slowly and carefully and put it on top of the additional idle Warré hive box. Some bees may fly out to see what's going on as you do this, but if you're slow and gentle and channel your inner tree, even if it's slightly swaying, most of the bees will be completely undisturbed.

To reduce the chances of squashing any curious bees, rest this top box down so that when it touches the box below it is rotated

45 degrees to it—so the two boxes' corners make an eight-point star. This allows the two boxes to touch at four points with the smallest area that allows effective stability with minimum bee squishing. Slowly rotate the top box to align with the box below, giving bees time to safely escape the scissoring wooden walls.

Carefully replace the top bars in the lower box of your bait hive, lift the stack of two idle Warré hive boxes you've just made, and put it on top of this lower box in the same way you will now always place one hive box on another—the "star turn" that should always top the bill.

Replace the roof and your idle Warré hive is now complete with bees that chose it. The insulating wool you added will allow them to survive the coming winter with less stored honey, which means that swarms which show up later in the year have a better chance. We might even be able to rewrite that rhyme:

"A swarm in May is worth a load of hay,
A swarm in June is worth a silver spoon,
But if they swarm in July, keep the sheep's wool close by,
If they swarm in August, they're brave but not thoughtless,
If they swarm in September, could be none to remember,
But never say 'never' . . ."

Some beekeepers are nervous about swarms because "you don't know what you're getting." They prefer to buy a nucleus or a queen from a bee breeder with a pedigree. But all honey bees are wild, like any animal whose breeding we can't fully control. Even when breeders instrumentally inseminate queens, how do they select the male drone bees they harvest for sperm? How do they select the queens? And how significant is their selection in comparison to the bees' own natural selection? Breeders strive to combine honey productivity with "docility" for the benefit and ease of the beekeeper, but these traits may well not be in the bees' best interests as is evidenced by "docile" queens sometimes becoming, like many of us, grumpier as they get older. And successful breeding for reliable increased honey production still proves elusive. It may be that in spite of their best efforts, bee breeders are farting uphill into a gale of evolutionary wind. There may not be so much difference between swarms and bees of "breeding"—like when new neighbors move in next door, none of us ever really know what we're getting.

But would you rather live next door to people who've been forcibly rehoused with no say in the matter, or neighbors who've chosen their new home because after checking out all the amenities and investigating lots of other places they all agree this one's the best? Unlike the bees in a nucleus you've purchased, the bees in your bait hive know exactly why they're there—some of the reasons we understand, some we can't, but those bees know them all.

And the most unexpected thing I've discovered about twenty thousand bees unanimously choosing to make their home in a box you've put out for them is that in that extraordinarily powerful and beautiful moment when they arrive with dazzling singularity of purpose, you don't just feel chosen . . . you feel blessed.

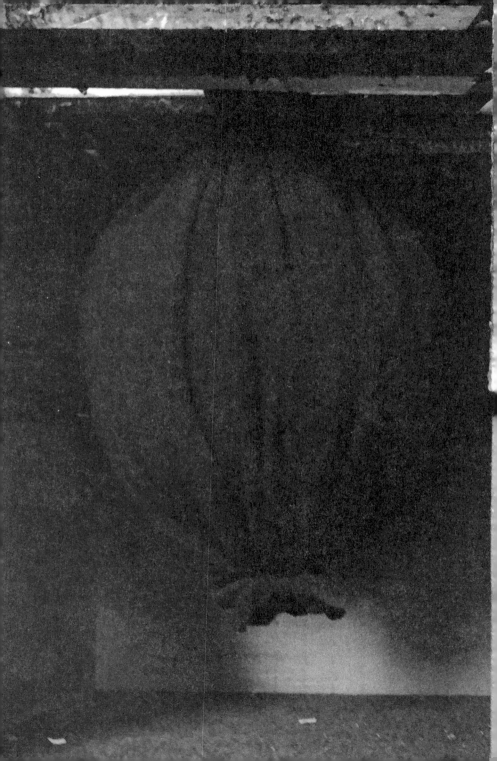

6

Sweet Secrets

Wherever your new bees have come from, they get to work inside your hive straight away. You, on the other hand, after the excitement of their arrival, now have an extended period of idleness: apart from a brief visit in the autumn to prevent mice from moving into your hive for the winter, there is nothing for you to do until spring next year.

Sadly this "nothing" almost certainly includes no honey—until autumn next year—because your bees probably won't have any to spare: they might need every drop they make to survive their first winter with you.

Forensic grammarians with sweet teeth will be chewing the conditional of the previous sentence—"almost certainly," "probably," and "might" are not absolute terms of embargo. So for their benefit, and for anyone vaguely interested in eating honey, let's take a closer look.

There are hundreds of different species of bees, but only one that makes such significant quantities of honey that it gets included in their name. So why do all those others shun the sweetness?

A familiar representative of that majority is the humble bumblebee. Their queen has much more modest aspirations in terms of a home for her colony than does the honey bee. Bumblebees recycle abandoned mouse nests for their homes: typically they have a volume of less than two quarts—twenty times less than the relatively palatial ten gallon tree cavity the queen honey bee needs for

OPPOSITE *Under wraps*

her colony—and much of those two quarts will still be filled with the bedding the mice installed to keep their litters of young warm. This bedding is a big part of the appeal of the mouse nest for the bumblebee queen, who also needs to maintain a constant comfortable temperature for her young. The mouse nest is big enough to accommodate a maximum population of about five hundred bumblebees, tiny compared to populations of more than sixty thousand that can be found in honey bee colonies.

This massive difference in scale begins in spring every year when the queen bumblebee comes out of hibernation with a retinue of none. She's the only bumble who survives the winter, a single mom emerging from the cold to search for a new nest. Once that's been found, she faces a multitasking juggle to begin a new family from scratch.

By herself she flies around collecting nectar and pollen, returning to the nest where she makes a single wax honeypot which she then fills with enough regurgitated nectar to see her through a couple of days of bad weather. She moistens pollen with her saliva to start it naturally fermenting into bee bread. Then, positioning herself near enough to honeypot and bread so she can feed herself without moving, she lays eggs, which she broods at the necessary constant temperature with her warm abdomen, like a bird. Of course, as any single mother will tell you, when the food runs out there's nobody else to go to the shops, so the queen bumblebee has to quickly nip out to some hopefully convenient local blossoms—while she's away her brood are cooling down, so she can't be gone for long enough to gather much more than hand-to-mouth supplies. Not glamorously regal, but impressive that her family can get so big from such humble single-parent beginnings.

By contrast the queen honey bee kicks spring off with an adult retinue of about five thousand bees ready and willing to resume the industrial process of nurturing her lineage. The moment spring flowers open for the business of nectar and pollen, thousands of honey bees are immediately available to fly out and gather it. Their queen is virtually incapable of feeding herself—her entourage deliver food to one end of her while she delivers eggs from the other. Not glamorously regal either, but impressive that with such support she can go on to lay as much as her own body weight in eggs in a day.

The advantage of the honey bee's population explosion at the earliest possible moment is that it allows the colony to sprint toward that critical mass of bees that allows it to divide in two, like a living cell, as early as possible in the year. And that gives both the swarm and the remaining colony the most time to gather the most food to give both the best chance of surviving the coming winter. And in a good year this can be swarms plural—a colony might be able to divide not just into two but three or four.

The cost is having those five thousand bees on the starting line, ready to go—alive. They all need to be warmly accommodated and fed throughout the winter. And the bill is paid in honey. In those winter months there is no nipping out to feed the five thousand because there isn't any food. Honey is liquid sunlight, saved for when flowers can't meet the bees' immediate energy requirements. Winter fuel to tide over the gap between the fading blooms of autumn and the explosion of spring blossom.

And if the honey bees haven't stored enough to see them through, their industrial model collapses and they die. All of them. Around one in five newly established natural colonies doesn't survive the winter. The same lethal outcome occurs even if the bees have worked successfully to store enough honey but someone removes so much from their larder that what remains is insufficient. Culpable bee slaughter. Extracting so much rent from your tenants that they starve to death is not a great long-term business model for landlords, and the honey we take in exchange for providing our bees with top-notch accommodation has to be what they can afford to spare in any given year—not necessarily what we want or expect. The bees work ceaselessly to expertly build and maintain a family in a changing environment—but they have no knowledge of how many slices of toast we may have buttered, poised, waiting for honey . . . and neither do the flowers that feed them.

Plants offer nectar to flying insects out of self-interest. Their goal is to have sex with as many other plants of their species as possible. The farther you go from a plant, the more potential sexual partners, but the plant can't go anywhere—it's rooted to the spot. Pollen, plants' dusty sexual currency, is light enough to be carried on the wind and, with luck, land on a suitable paramour. But the odds against hooking up with "the one" border on that fairy-tale ratio of how many frogs you need to kiss before you find a prince. Far better to stow away some pollen on an insect that's deliberately flying off to other plants where it can do some precision intermingling for you. And no requirement for random frog-kissing. Such a good deal for the plants they offer free aviation fuel to encourage any passing insect to drop in for a drink and a dusting.

Nectar is a liquid carbohydrate carefully designed for flying insects. The energy it contains is as exactly balanced to their demands as the gas we pump is matched to the needs of our car engines. We don't want to fill our tanks with half-strength gas that will give us less range, speed, and power, but our hydrocarbon manufacturers don't want to waste valuable energy concentrating more than we need into overproof gas. It costs plants energy to produce the nectar they freely give away, so they try to get the maximum value out of pollinating insects by targeting their supply. Wantonly pumping it out would take them back toward puckering up for frogs again, so they control its flow to coincide with weather conditions that favor flight. Plants that are pollinated in darkness by night-flying insects won't release nectar during the daytime and vice versa. Extremes of temperature that discourage flight reduce nectar flow. Sometimes you can see insects pollinating a carpet of blossom in a formation like cattle strip grazing: ignoring one swath of blooms but devouring their identical-looking neighbors—like farmers ensuring that their fields are grazed evenly, plants can control the flow of nectar to encourage the insects to give every blossom a fair chance of a good fertilizing encounter.

This all-you-can-eat but no-more-than-you-need fuel arrangement works well for most insect pollinators. But honey bees have an agenda that even bumblebees don't share—the feeding of the five thousand when the blossoms aren't there—so honey bees will harvest nectar beyond their immediate requirements and take it back to the hive to store in huge quantities.

But storage requires space. As each cell in the comb becomes vacant, there are competing pressures to fill it: either with a baby that can grow into a nectar collector, or with pollen or winter fuel. But there is not enough energy in the raw nectar to successfully balance these choices in the space available, and not enough time or energy to build more empty combs. If the bees were to simply store the nectar, there wouldn't be enough room for their population to survive the winter consuming it.

So to save space the bees remove the constituent of the nectar that yields no useful energy—the water. Depending on the time of year, the time of day, and the plant, nectar consists of about 80 percent water and 20 percent sugar. Foraging honey bees drink nectar from flowers to power their flight, but they also gather surplus nectar in special crops—internal pouches they fill through their mouths, but distinct and separate from their stomachs. A forager doesn't digest the nectar in her crop but starts preparing it for storage in two ways: absorbing some water through its walls, and adding the enzyme *invertase*, which starts breaking the sugar down into fructose and glucose as she flies back to the hive, often carrying nearly half her own body weight of nectar in her crop.

There she will hook up with a housekeeping bee near an empty cell. These two bees will lock their mouthparts together and the forager regurgitate her load of nectar into the nectar crop of the housekeeper before flying out to collect more. The housekeeper then carries on the process of adding enzymes and absorbing water through the walls of her nectar crop but also regurgitates a big droplet of the nectar to hang from her mouthparts, allowing water to evaporate from its maximal surface area. She'll swallow it back into her nectar crop, mix it with the rest, and regurgitate another droplet to hang out to dry. Once this load of partially dried nectar is fully enzymed up, she'll regurgitate it into a cell of honeycomb, ready to receive the next load of fresh, wet nectar from another forager.

Between foragers she'll use her wings to fan the nectar she has put into a cell to help it dry, simultaneously splashing it with her long tongue to increase the surface area where all the evaporation takes place. Eventually the water content of the nectar is reduced from 80 percent down to 18 percent and at that point the drying stops because the nectar has become honey. Again, this is not a number chosen by that conspiracy of Illuminati numerologists. At 18 percent water, the honey now takes up less than half the space it occupied as nectar. But successfully solving the problem of limited comb availability is no longer the only factor: sugar storage is not just about space—it's also about time. There's no point in beautifully accommodating enough energy if it's not going to keep for as long as you need it.

It's not just you, me, and the bees that find honey irresistible. But at 18 percent water, the concentration of the sugars in the honey is *just* high enough to be toxic to organisms that thrive in a wetter solution—yeast, bacteria, and fungi adore damp sugar, but like all of us who believe that you can't get too much of a good thing, it's only because we haven't actually experienced too much yet. All poison is in the dose, and the bees increase the dose of sugar in their honey to preserve it from the ravages of microorganisms that can't survive its elevated sweetness.

Kept dry, it doesn't go off. Two-thousand-year-old honey found in the burial chambers of ancient Egyptian pyramids is still delicious. And I'd hazard a guess that honeycomb buried intact for ten thousand years in the stone grain store of a Mesopotamian farmer, who was one of the first humans to cultivate wheat, would contain honey that would taste as good on your buttered toast today as it did when toast was invented.

Honey is hygroscopic—it will absorb water from the surrounding atmosphere, so like those ancient Mesopotamian bees, your bees will seal the honey-filled wax cell with a cap of waterproof wax to prevent all their drying work from being undone. There's very little margin for error here: the drying process demands so much energy that the bees won't usually waste any drying it out more than is absolutely necessary.

Jam makers who practice the same brinksmanship sometimes find the contents of their jars start to bubble in alcoholic fermentation when the lid's been left off too long and sufficient water from damp air has been absorbed by the jam to make it wet enough to be hospitable to yeast—spoiling the jam and wasting all the energy used to boil off the water in the fruit.

And though any thought of jam today may be a poor substitute for the honey you can't have from your hive yet, the way we make jam can illustrate how the idle Warré hive at least speeds up the honey-making process a little.

Unlike jam makers, housekeeping bees don't have a hot plate or copper Maslin pan they can use to speed up removing the water from nectar by boiling it off. Instead they take advantage of all the heat being generated to keep the babies warm by positioning themselves on the comb directly above the nursery area. So when they regurgitate that droplet of nectar out into their mouthparts, it's hanging in a current of warm air rising upward, and both the heat and the motion of the air help evaporate water from the droplet. Try it for yourself: quickly lick the backs of both your hands and blow gently on the saliva on one of them—which hand dries first? The closer you put your mouth when you're blowing, the warmer the air that hits the saliva and the quicker it evaporates. The surface of the partially dried nectar will similarly evaporate more water more quickly in a warm draft.

Canny housekeeper bees use baby heat wherever they live, but the idle Warré hive helps them more than most: and it's those sheep again. If you're a housekeeping bee working bang in the middle of your hive, you'll be operating at a familiar, beautifully regulated temperature of 95 degrees Fahrenheit in pretty much whatever design of hive you find yourself. But there are hundreds of housekeeping bees working away, and many of them are positioned close to the walls of the hive. Whenever it's colder than 95 degrees Fahrenheit outside the hive, in an uninsulated hive the temperature of the inside of the walls will be closer to the outside temperature—colder. The housekeeping bees working next to the walls will find themselves in a current of nursery air that is cooler than the current of nursery air in the center of the hive. And this not only means they get less heat to help with drying the nectar: because cooler air rises more slowly than warmer air, the colder walls of an uninsulated hive also slow down the speed of the draft of air moving upward, reducing the evaporating energy of its motion.

Repeat the hand-licking experiment but now blow *much* harder, and keep blowing—the increased speed of evaporation from the faster current of air you create will dry your hand even faster, hopefully before the hyperventilation kicks in.

In a tree cavity or an equivalently insulated idle Warré hive, housekeeping bees evaporating nectar near the walls will be working in a hotter, faster draft of air than will their counterparts in an uninsulated hive. This subtle effect will persist over the hours, days, weeks, and months of the summer and allow your bees to turn nectar into honey more quickly. And if there is an abundance of nectar in the plants outside the hive, by condensing it into honey more quickly the housekeeping bees can make more cells available for more babies to become more foraging bees to gather more of that abundant nectar to deliver to more housekeeping bees . . . and the colony makes more honey, which makes a surplus more likely. And that's the sweet spot for our buttered toast. But a surplus isn't just about how much honey the bees make in the summer—it also depends on how much they consume over the winter.

With flowers gone, seed pods rattling in icy winds, and bees huddled together eating honey for warmth in a cluster around their queen in the hive, this is the time for another woolly experiment, just to satisfy my curiosity.

I want to know just how much energy the wool-wrapped idle Warré hive might save the bees compared to an uninsulated hive. In winter the bees don't waste energy keeping the whole hive warm; they concentrate on the cluster they gather into to conserve each other's heat. This cluster is as spherical as possible so they can be their own best insulators, and it's this I want to emulate. So I find an unloved, deflated football abandoned in the garden and use it to form two spheres of chicken wire. Around the corner on Portobello Road Market there is a stall that sells pure cashmere jumpers previously loved by moths for $5. I cut one into two equal parts, briefly revive the anti-consumerism craft of darning to fix the holes, and then slip each part onto a sphere, gathering it together at the top and bottom like a giant woolly sweet wrapper. This cashmere wrapper mimics insulating bees and delineates the boundary of their "cluster." It's the energy used to keep the inside of these cashmere enclosures at bee temperature that I want to measure, because that will give an indication of how much honey each needs to consume.

Details of the experimental methods can be found in appendix 5, but daily monitoring revealed that in order to keep the "clusters" at bee temperature over the winter, the one in the uninsulated hive consumes more energy than the one in the idle Warré hive, not by 5 percent, not 10 percent, but 50 percent. Half as much again. Your idle Warré hive uses 33 percent less energy.

So there you have it—insulate your hive in sheepswool and all that 33 percent energy saving will be honey that isn't consumed by the bees but is instead added to the surplus that is yours. The more wool you add, the more honey you get. Basic accounting.

Way too basic. Even I don't need professorial help to know that this experiment—indicative though it is—is a gross oversimplification of what goes on in a real cluster full of bees.

Those combs hanging full of the honey that the cluster is slowly working its way through become better insulators as they're emptied and become full of air.[3] So every drop of honey the bees consume to produce heat increases the insulation around the cluster. By the end of the winter, although their honey supply is low, the bees that consumed it are slightly better insulated against the cold.

But this increase in insulation is offset by a loss of thermal mass: the honey the bees have collected stores heat very effectively, and as it is consumed this stored heat diminishes.

The bees themselves are both heaters and insulation by turn. Like emperor penguins in the Antarctic who huddle together to brave the blizzards, the ones on the outside of the group insulate those on the inside, swapping places in rotation before they freeze to death. The "mantle" bees on the outer extremity of their cluster perform an additional life-supporting role: it's not just food that living things need to survive, it's water. Thirsty penguins only need to face into the wind and open their beaks to get a drink from an Antarctic blizzard. Bees spent all summer removing water from their honey, but now it's their only source in the hive. And flying outside in cold weather to collect more would be not only potentially fatal but also extraordinarily energy depleting—just try carrying two full buckets of water more than a hundred yards—and then imagine how out of breath you'd be if you had to fly that far with them.

So the bees have developed a water-recycling system. The consumption of honey not only releases energy, it produces carbon dioxide and water vapor that, as do we, the bees breathe out. This warm bee breath passes though the cluster, and when it comes into contact with the chilled mantle bees, hanging on for dear life at the cold face, the water vapor condenses into droplets on their bodies like dew,

---

3   Low thermal conductivity means good insulation: Honey's is 0.5 W/(mK); air's is much lower at 0.02 W(mK).

and they crawl back inside the warm cluster where it's mixed back into the honey.

Lord Kelvin would be the first to point out that the mantle bees are conserving not only water but also heat: the laws of entropy decree that when a vapor condenses into a liquid, heat will flow from the hotter to the cooler—from the warm bee breath into the cold mantle bee—so less heat escapes from the cluster.

But what if honey production isn't the bees' main priority? What if honey is just a means to an end—to allow the bees to stay alive long enough to ensure the survival of their DNA? Like all living things? In some beekeeping circles it's almost heretical to say it, but by this logic the primary produce of a colony of bees is not honey but healthy drones—the male bees who carry the DNA of their queen out into the world and into the future.

I can relate to the resistance to this drone-centric view for two reasons: first, honey is delicious; and second, as someone whose immediate DNA was forged in the Presbyterian Northeast of Scot-

Drones out on the tiles—they're not all posing in aviator sunglasses; unlike female worker bees the drones' eyes touch together on top of their heads.

land, I'm lumbered with a latent respect for the Protestant Work Ethic. Drones are lazy bastards. They don't lift a finger: they do no work, they drink all day, never pay for any of it, they can't sting, and they're fixated on sex.

I don't know how the female worker bees who build the comb, raise the kids, and fetch all the nectar to make all the honey put up with them. The idea that this exquisitely organized, dedicated, intelligent, and industrious sisterhood could be doing all that work so a bunch of big boozy slobs can get a shag sticks in my Scottish craw.

Even when we discover that the drones might help a tiny bit with the heating, I imagine that, swollen with scrounged nectar, as they press their big, furry, hot abdomens down on the comb to warm the brood, their intensely competitive masculine nature would have them insensitively exchanging risqué locker-room banter with fellow gut-dragging drones in front of the children. But maybe that's just me—perhaps they're singing lullabies . . .

And it seems to work for the bees, even for the female workers who have all collectively delegated their fertility to their mother the queen. If that queen dies and they cannot replace her, the colony is doomed. And nothing concentrates the hive mind like the proximity of death. In extremis, priorities reveal themselves: when a colony can't replace its dead queen, the bees don't start desperately increasing honey production—in these circumstances a worker bee can start laying unfertilized eggs containing the DNA she shares with her late mother, and the colony dedicates all its resources to raising them to adulthood before its demise. They will all hatch as male drones who will fly from the hive like a last, desperate genetic message in a bottle, thrown out to sea, in hope, from a sinking ship.

Whereas female workers are always scented with their mother's individual perfume and are not welcome in any other hive—they just smell wrong—male drones are welcome in any hive they choose to visit. They'll be given food and drink by its hardworking occupants who seem to expect nothing immediate in return—the drones are free to leave and go where they will without paying. Because all the colonies in all the hives understand that it is good not only to selectively share all their DNA but to share it as widely as possible. So whilst the drones might think they're on the most brilliant pub crawl with free drink and friendly female bar staff, it's all set up to allow them to fly farther from their mother, carrying her DNA. A female worker bee can fly as much as five miles to find food, but she has to return to the only hive where she can survive. A drone can hopscotch hives for dozens of miles before he finally finds the love of his life and delivers, for Mom.

If using the absolute minimum of honey in the winter were the bees' top priority, all five thousand overwintering bees would hibernate like bumblebees. Hibernation involves radically changing their metabolism so they enter a state of sustained near-death—everything slows down as near to total shutdown as possible so they use the absolute minimum amount of energy to stay alive. Honey bees don't hibernate because although the energy savings are very significant, the radical metabolic shifts involved aren't quick. Waking up from deep hibernation—surfacing from near death to fully alive—takes time. And the whole point of the five thousand overwintering bees is to be ready to steal a march and crack into those first blooms of spring the second they arrive. Even if it's just for a two-day weather window before more snow—which groggy hibernating bees might wake up just in time to miss, and then have to go all the way back down into hibernation again. So although not hibernating costs the honey bees huge amounts of honey—most of their winter stores—they seem to be willing to pay that price for increased speed and flexibility of deployment in the spring.

Insulating your idle Warré hive throughout the year will allow the bees to keep more of the energy they gather from the outside world inside your hive. But there is no guarantee that the bees will use all that energy you have saved for them to make honey for you.

They might well decide to use some of that energy to make more bees because that is their top priority. Of course, those extra bees will need extra honey stored for them. But the colony will factor that in too and allocate energy accordingly.

So we need to be cautious when we estimate the surplus of honey we can safely take from the bees. And though a vigorous swarm that moves into your bait hive early in a very good year for blossom might produce a small surplus of honey, I'd leave it for the bees. Imagine if you carefully calculated what they would need, removed the surplus in the autumn, and then went to visit your hive in the spring to find all the bees dead from starvation—desiccated corpses, some with their heads right down inside empty cells of comb where they died searching and scraping in vain for any last traces of honey. A traditional skep beekeeper, who routinely killed all their bees to harvest for honey, would look at this accidental annihilation and ask, "Why didn't you just kill them before they ate all the honey?"

The same lack of a crystal ball that prevents bees from predicting the future in summer continues through the winter. In summer this means they simply work as hard as they can to gather and store as much food as possible to sustain the maximal cluster over the winter. Pedal to the metal. This cluster cannot know how long the winter will last, so the bees simply, frugally consume as little as they can to stay alive as efficiently as possible. There's no evidence of wild, profligate, bacchanalian cluster parties spilling honey with reckless abandon.

It can be very frustrating having to allow others to set the pace for something you want immediately. When I'm flying, traveling in an airplane that starts experiencing some turbulence, my adrenaline and cortisol response is every bit as good as the next passenger's, and we all want the impotent hell of our shared fear to stop now. Alcohol doesn't help, but I've found that the most calming, rational response is to honestly consider that whilst my every instinct is to take action, right now, to fix this, would I be able to do anything better than the pilot?

And in our pursuit of honey, unless we think we can gather nectar and process it for the long-term good of the colony more effectively than the bees, maybe it's best we stay out of the hive in the same way as we should stay out of the cockpit.

Because outside the hive, there are things we can do to help.

7

Bee Observant

Since they stopped working the land, many of my family have become passionate surgeons: plowshares to scalpels. But if we consider the hive as a womb where bees are protectively nurtured from egg to delivery into the world, opening it up to have a look inside to confirm that all's well might be the equivalent of performing surgery on a healthy pregnant woman every time she has a routine check-up, without even asking her how she feels. Even my family wouldn't scrub up for that. The bees can't tell us how things are going inside their hive even if they wanted to, but there is much we can see from their behavior at the entrance that reveals the state of the health of the colony within and might avoid wholly unnecessary surgical intervention.

My grandfather was performing emergency surgery in military field hospitals in France during the First World War at the same time as Abbé Émile Warré introduced the first version of his People's Hive: Warré based the design of its roof structure on the military tents that had come to his diocese in the Somme. But as the war ended and Warré continued to refine the design of those most simple and basic square wooden boxes you have used to build your idle Warré hive, his fellow countryman Dr. Adolphe Pinard, a Parisian obstetrician, developed the most simple and basic round wooden ear trumpet—pressed against a pregnant woman's abdomen, the Pinard stethoscope allows her baby's heartbeat to be monitored. It's still preferred by many midwives today because, without electricity or

OPPOSITE The Friend of the Bees *by Hans Thoma (1839–1924)*

microprocessors, it can be used to locate the baby's position more accurately than a Doppler fetal monitor. Both Warré's square and Pinard's round structures concentrate the thing we're interested in at a small opening in their wood, but we "see" the heartbeat of the hive on the back legs of the bees flying in at the entrance via the pollen they carry in.

Even at the busiest times at the height of the summer, not every bee will be carrying pollen into the hive; some will be fully laden with a cargo of nectar, invisible inside their abdomens. Bees need both, but the big difference between the pollen and nectar flowing into the hive is not whether we can see it or not—it's that pollen contains nitrogen, the invisible element that makes up 80 percent of the air we breathe.

At a molecular level, nectar is a sugary carbohydrate made up of carbon, hydrogen, and oxygen used as fuel to power life. Like us, bees can concentrate and store excess fuel by converting it to fat inside their bodies, and outside their bodies they can remove water to concentrate and store it as honey. But it's still fuel.

Car drivers know that though gasoline or diesel is essential to make their cars go, it's really hard to make a new door panel, or exhaust pipe, or even a tire from either type of fuel. To be fair to budding alchemists, there's probably a healthier margin turning lead into gold, but however you juggle its bits around, ultimately fuel can't make anything but heat and exhaust gases: energy, carbon dioxide, and water.

In living things, if you want to make new life or repair old, fuel isn't enough. Into that sugary carbohydrate mix of carbon, hydrogen, and oxygen molecules you need to somehow add nitrogen: then you've married into the nitrogenous family of amino acids and proteins, and you don't just have a zest for living that comes with fuel—you've made the "stuff" of life, its substance and structure. Every cell of every living being is replete with proteins.

But getting together with nitrogen is really tricky: it prefers to stick to itself—literally. Even though we inhale four times as much nitrogen as oxygen in every breath we take, we can't absorb any of it to make protein because in our lungs, nitrogen molecules irresistibly cling to each other and are exhaled with their grip undiminished.

Plants are no better than us at absorbing nitrogen from the air, and they can only grow because, in one of the most important bio-chemical processes for life on earth, bacteria in soil can "fix" airborne nitrogen so it becomes available to plants' roots. And though there is negligible protein in their sugary nectar, the pollen that plants make available to the bees is protein-rich—it contains more amino acids and protein than the same weight of beef, eggs, or cheese.

Lots of pollen going into the hive equals lots of protein to make baby bees. Indoors a healthy queen will respond to the protein sup-ply and lay eggs: she's converting plant protein into baby bee protein, and those growing babies will continue the conversion, feeding on the protein in the fermented pollen being fed to them by nurse bees. The population of the hive will be increasing and the colony will be flourishing. Like blood cells flowing from and returning to the heart, the foraging bees leaving and returning to the hive with pollen are the most fundamental sign of life and health in the colony.

So if you're considering opening up your hive to have a good look inside, but as you approach it you see dozens of bees flying in with pollen every minute—stop! All is almost certainly well! Let them be.

If you've placed your hive where you can see it regularly, you will start to become familiar with its individual characteristics. When you have a few hives side by side, you'll notice that the behavior of the bees at each of the entrances is subtly different: some colonies' foragers fly in and out so decisively they barely touch the landing board; oth-ers seem to meander around a little before eventually committing to fly off or hover cautiously before landing on return. Some entrances have a prominent crowd of high-visibility security guards, eyeballing all that come and go; others appear to be undefended, until a wasp or a forager that's not a resident gets too close and is suddenly bundled away by a guard that's powered out of the hive to see the intruder off with extreme prejudice. Some colonies often have a bee or two with their backsides pointing out of the entrance, fanning their Nasonov glands to helpfully identify their hive with its particular "over here!" scent to guide their sisters home, while other colonies don't feel the need. Each colony has a nature as individual as your children or your friends. And the first time you really notice this is when its behavior changes—when you instinctively know that "something's up" before you can even describe what's different, and then start consciously working out the symptoms of the shift in temperament.

Sometimes the activity at the hive entrance changes because the gentle fluctuation in the amount of pollen visible on the hind legs of returning foragers has become dramatic. This is rarely because it suddenly increases—a rapid doubling in the number of foraging bees is extraordinarily rare, and the sight of twice the pollen being flown in would gladden most beekeepers' hearts. But if the pollen flow is abruptly halved or worse, this reduction will affect the babies' food supply and consequently reduce the number of eggs the queen will lay. Usually this dearth is temporary and caused by a blossom "gap"—the varieties of flowering plants in your locale bloom at different times, and right now there may be very few in bloom: there's just not much food out there for the foragers to find until the next species flowers.

Whatever the cause of the blossom gap your colony is suffering, you can help fill it for next year by planting bee-friendly flowers that will be open for nectar and pollen when your bees need them. There are all kinds of lists, recommendations, and claimed flowering periods,[4] but nothing compares to finding a plant not too far away from you whose blossom is covered in bees when you need it to be. When you find one, you can of course buy the same variety as a mature plant from a garden center, but it's also worth asking the owner if you can take cuttings of their bee-laden plant: the owner will be able to tell from the way you look at the bees that your explanation is sin-

4   https://www.swallowtailgardenseeds.com/tips_lists/bee-friendly-plants.html

cere, and you will remember that individual's generosity every time the blooms grow from their plant.

The other advantage of a small cutting that has taken root is that it doesn't seem such a big ask for one of your neighbors to plant it in her garden. She may already have blossoms when your bees need them, but they may not be bee-friendly blossoms: buddleia may be great for butterflies and bumblebees with their long tongues, but honey bees' shorter tongues can't reach deep enough down into the bloom to get at the nectar. And for the same reason, whereas white clover is a staple for honey bees, the deeper blooms of red clover only tantalize them. Finding a small spot for your little cutting might be the germ of an understanding that could transform a neighbor's garden into an uninterrupted feast for the bees.

Increasing the proximity of bee-friendly plants flowering at the right time, however precisely targeted, might seem merely a morsel of feeding in a gale of want: to enable a colony to thrive over the period of a blossom gap, you might need to plant agricultural quantities of food, not just persuade a neighbor with a window box to accommodate one plant, however delicious. But for every colony there is a tipping point for its survival, and when times are hard, such as in a prolonged period of drought, a daily-watered window box might just make the difference between life and death.

Sometimes in the middle of winter when I'm sleepily hand-grinding beans for the first coffee of the day, the approaching sound of the dawn garbage truck is more effective than the shot of caffeine it defers: I rush around gathering empty bottles and more for the recycling bag, curse the entire family who've stuffed the kitchen bin so full that its bag is harder to extract than an impacted wisdom tooth, and tear outside into the freezing cold to catch the fluorescent trash collectors with my latest offerings before they pass by. Usually barefoot, even in snow, with a poorly tied dressing gown covering very little, I am ridiculous, shivering proof of the strength of our urge to expel rubbish before it begins to fester.

Bees have no bins to put out, but like us they have bowels and they scrupulously void their waste outside the hive. In winter they'll hold on for the warmest, sunniest day, but when you've got to go, you don't want to be hanging about courting fatal exposure to the freezing cold. So the bees will leave the warmth and kinship of their winter cluster inside the hive for a quick "cleansing flight"—the dynamic opposite of a "comfort stop," but delivering the same relief.

And this has consequences for gardeners. Not from windfalls of bee manure piling up near the hive—their droppings are tiny—but if you're up for planting bee-friendly plants to bloom in the summer, consider the winter menu too.

I've never stood semi-naked out in a frost, relieved to be watching the trash collectors taking my rubbish away, and thought, "Now I'm out, I might as well walk barefoot to a huge supermarket miles away and get in a week's food shopping." But if as the grumbling refuse truck departed I began to pick up the scent of warm, sweet cinnamon buns coming from a little coffee kiosk that had magically appeared just across the road, I might think that I wouldn't get *very* much colder or look ridiculous for *very* much longer if I popped over to check out the source of this delicious smell. And imagine if in recognition of my enthusiasm the barista in the kiosk offered me a cinnamon bun for free! And then pointed to a sign that said, "Free coffee with every bun!" Even my frost-nibbled feet would be happy so long as he ground the beans faster than I do.

Winter-flowering plants can be the real-world equivalent of the fantasy pop-up coffee kiosk for the bees, but the crucial thing is they need to be close by, so a bee that's just nipped out for a number two without even putting its dressing gown on can get a quick free meal with minimal risk of hypothermia. Gardeners can help ensure a good population of bees to pollinate the spring blossoms of their autumn crops by planting winter-flowering heathers, aconite, early crocuses, primrose, and lungwort near to the bees' hive.

Even better than a coffee kiosk is a festival catering truck, and whilst mahonia is a shrub that offers thousands of bees a feast of winter plenty, my personal favorite is winter honeysuckle: cream flowers dangling golden pollen burst like fireworks of sunlight out of bare wood stems in a fecundity of nourishment from December to March, but it's the power of its heavenly scent that makes it something we can truly share with the bees.

The twig I planted in my front garden is now a six-foot-tall bush that reaches out into the pavement, and in the winter I sometimes see passersby behaving like bees—they are walking purposefully along the street, agendas unknowable, and then something reaches into their minds and they waver: they look around, trying to make sense of a scent so powerful and sweet and which the cold of winter has made so unexpected. Can it really come from this single plant? They stare at the blooms of the honeysuckle in melting disbelief. Gorging on the olfactory feast that *Lonicera fragrantissima* invisibly provides, they see dozens of bees collecting nectar and pollen, busy in the middle of winter. A mesmerizing, life-affirming conundrum.

Like me you may be a very poor and inexperienced small-scale gardener, but the honey bee is a super-pollinator and will leverage, upscale, and maximize any seed capital you invest. She is by no means the only pollinator on the block, but there's something special about the myriad of different colors of pollen you can see on a honey bee's hind legs in summer as she disappears inside your hive: they're evidence of a diversity of plants currently in bloom. For the price of a pint of beer in a London pub, you can get a set of pollen identification cards, which may not deliver the instant gratification of a similar-sized credit card, but they'll never leave you wishing you'd taken a pair of scissors to them. Pollen identification cards will tell you the flowers your bees have been visiting based on the specific color of their pollen. You might be amazed at some of the species your bees have found in abundance within three miles of your hive, plants that you've never seen.

But being able to identify the flowers from the pollen the bees are carrying is only possible because of the trait bees learned that turned them into super-pollinators. When a honey bee visits a flower, the plant has the best sex—and the bee's not even trying. The pollen-foraging bee is focused on gathering protein as efficiently as possible. She's on a supermarket dash that costs her a lot of aviation fuel. Everything's free in the supermarket of blossom but only till sunset, so the bee's trying to fill her cart with food and get back home as quickly as possible, so she can make more trips to maximize the amount of sustenance for her family.

Different plants have different pollen grains: they're not only different colors, they're also different shapes and sizes. Bees have learned that the grains of pollen of a particular plant are always pretty much the same shape and size. But much more importantly they've also learned that in the supermarket dash, you can fit more pollen into your cart if you exclusively stick to grains that are the same shape and size: they pack in better. When you're in a hurry, a cart full of cans of beans, followed by a cart full of cabbages, holds more of both than two carts with a higgledy-piggledy mixture. So when a bee comes to fill the basket of hairs on the back of each of her hind legs, she sticks to the same pollen from the same plant. A bee will forage from different plants during the day, but on each flight she will exclusively collect pollen from a single species of plant.

For the bee, this pick algorithm maximizes her pollen load per flight, which allows her to save aviation fuel while delivering more food for the babies more quickly, so she can make more flights per day.

For the plant, it means that the canny honey bee that arrives to pollinate it always comes covered in pollen from plants it can have sex with. Stripping the pollen off the anther and stuffing it into baskets of hairs on the back of her legs is a dusty business, and however efficient the bee, she'll spill some loose grains—all part of the plant's plan. When she takes off for the next blossom, the second her legs break contact with the plant she's no longer electrically grounded, and her beating wings build up a static charge in her body that attracts loose grains of pollen like a magnet picks up iron filings. Electrically bristling with pollen grains during her flight, as the bee lands on the next flower, the static charge maintained while she was flying discharges to earth, forcing the grains of pollen that had clung to her body to fly out and make contact with the plant. For the plant to have sex, the right kind of pollen has to make contact with its stigma, and this electrical explosion increases the chances of this happening even if the bee doesn't directly brush up against it.

Most pollinators don't deliver with this precision: for the majority of other pollinating insects, pollen's all the same, they're not fussy about which plant it comes from—food's food—so for a plant, most pollinators have an increased chance of bringing random pollen from random plants of different species it can't have sex with. Nothing for its stigma to get excited about: any pollen that lands on its surface is immediately chemically interrogated by the stigma, which will ignore grains not only from another species but also from itself—the basis of this whole process of reproduction is to increase genetic diversity, not generate an incestuous selfie.

Every grain of pollen the honey bee arrives with is viable. She might well bring sexually interesting pollen from a potential mate three miles away that has all kinds of alluringly different attractions: and when that kind of pollen grain lands on its surface, the stigma purrs "Hello, gorgeous . . ." by releasing a chemical that triggers the grain to grow a microscopic pollen tube. Like a root sprouting from a seed, but on a much smaller scale, this pollen tube is a single cell that penetrates the stigma, growing down through the style to reach the ovary where it delivers the male sperm from the pollen grain to fertilize the female egg in the plant, and a seed is born. In some plants this single-cell pollen tube has to grow to twelve inches in length to get where it needs to go once it's been selected as an appropriate mate by the stigma, and it can do it at an astonishing speed: an inch every two hours is extraordinary compared to the rest of the plant, and such rapid growth could not be achieved without lots of protein—the greater the distance between the stigma and the ovary in a plant, the more protein its pollen needs to construct the tube to connect the two. And it's this high density of protein fueling the growth of the pollen tubes that also enables the rapid growth of population in your bee colony when the bees eat it.

Long-distance precision pollination by honey bees increases the genetic diversity and resilience of all the plants in the area of the hive. There will be more, healthier leaves, seeds, fruits, and berries from those plants to feed more and healthier animals, right up to the top of our food chain: us.

The other benefit to the honey bee of sticking to a single species of plant on each foraging trip is that all the plants on its flight path are likely to be producing nectar at the same time, responding with species synchronicity to the changing environment. This means the bee is able to more reliably top up with fuel at every bloom, giving her more flexibility to safely venture farther from the hive in the knowledge that she's less likely to run out of fuel on the way home.

And when she gets home she turns super-pollination into a social movement: she tells other bees where to go to get to the source of pollen she's just come from. And her social media app of choice is Waggle Dance.

It took forty-six years for Karl von Frisch's decoding of the meaning of the bees' waggle dance to earn him a Nobel Prize. In only a few seconds a bee can follow another bee dancing on the comb in a figure eight, waggling her abdomen, and learn the direction and distance from the hive to flowers providing nectar and pollen—the more rapid the waggle, the better the food source, and the more widely the good vibrations resonate through the World Wax Web of the comb. Like learning the coordinates of flight navigation by joining in a country and western line dance, it's easier to copy the steps correctly if the dance floor is perfectly level. Or in the bees' case, perfectly vertical. Why should this matter?

When the bees fly to flowers they've heard about from a waggle dance, they navigate by the position of the sun: in addition to the familiar two big compound eyes they use to see, bees have three tiny simple eyes. Although these "ocelli" can only detect light or dark, because they are arranged in a triangle on the bee's head and can detect polarized light, the bee can always triangulate the position of the sun when it's in the sky, even when it's hidden from our eyes by clouds. All flight directions in the waggle dance are given relative to the current position of the sun so bees can fly out of the hive and head in the right direction. But close your eyes for a moment and imagine you're a bee inside the hive. Dark, isn't it? Even on the sunniest day very little light gets though the entrance. The interior of the hive is almost pitch-black—the bees can't copy the waggle dance by sight, they have to do it by touch and feel.

If the waggle dance is a treasure map with the current position of the sun taking the place of magnetic north, how do the follower bees dancing in the dark of the hive know which way round the map is when they can't see the sun? How does the waggle dancer know that when she dances "Head 45 degrees to the right" her followers won't protest, "45 degrees to the right of what? We can't see the sun—we can't even see you!" What common data does her group of followers all possess that the waggle dancer can use to accurately communicate her direction of travel to all of them simultaneously?

The only way is up.

All the flight directions are given relative to straight up. Just as we can tell if we're vertical with our eyes shut, the dancer and all her followers know the feeling when gravity makes their abdomens hang as low as they go—when their bodies are vertical. And dancing on the gravity-assisted, perfectly vertical comb they built, they now have an accurate shared reference to follow a waggle dancer dancing "45 degrees to the right of straight up." The treasure map orientated, the bees fly out of the hive, translate their dark gravitational bearings to a bright "45 degrees to the right of the sun," and they know where they're going.

But imagine rehearsing a line dance routine, learning the steps by following a choreographer on a dance floor that was sloped, undulating, or uneven. You might be able to learn to perform the moves so you look like everyone else by compensating for your irregular section of dance floor, but what if you then had to perform that routine in front of an audience on a perfectly level dance floor? You might well be thrown by the difference in surfaces—you can't immediately uncompensate your moves for the level surface, and now your muscle memory could make you diverge from the choreographer's instructions even though you think you're following them. Any error in the verticality of the comb on which the bees learn the waggle dance will be carried forward when they "perform" the foraging flight to the dance's bearings relative to the sun. Such misleading information could critically affect the bees' ability to gather food.

But as any celestial navigator will tell you, knowing the position of the sun is only half the story: it's pretty reliable at telling you your latitude, but because the earth is always rotating, you can only work out your longitude—those north/south meridians that delineate the planet like segments of an orange—if you also know *when* you're looking at the sun. You need an accurate clock, and in 1714 the British Board of Longitude in Greenwich recognized this vital need for navigational precision by offering a prize of $25,000 (worth $20,000,000 today) to anyone who could make a clock that could survive the rigors of ocean voyaging and tell the right time.

John Harris made the winning timepiece by hand but only got his prize money in 1765 after King George III personally intervened to nudge the reluctant sponsors to cough up in full. Harris's clock allowed you to reliably calculate where you were like this: before you left on your voyage you would synchronize your clock to the master clock in Greenwich that would tell you Greenwich Mean Time; while at sea you measure the height of the sun above the watery horizon until you know the moment it's achieved its zenith—solar noon; you look at your onboard clock and see how much your current solar noon differs from the solar noon back in Greenwich—if the earth were an orange with 360 segments, every four minutes of difference in time you're one segment away from Greenwich, 1 degree of longitude. And if your current solar noon is earlier than Greenwich's, you're eastward; if later, westward.

Bees have been successfully navigating for fifteen million years without ever consulting John Harris's clock: circadian rhythms so perfectly attuned to a sun they can always position give them internal body clocks that are accurate beyond the endeavor of Harris's cogs and springs. Instead of a mechanistic tick and tock subdividing time and space into components to be measured, compared, and plotted, bees feel navigation like we feel the breeze: as they fly outside the hive, time, space, and position are in an evolving, fluid relationship where they are continuously being understood relative to one another. Bees just know what time it is and where they are. Like mindful GPS.

But for the first half of their adult lives bees live inside the darkness of the hive—how can their circadian rhythms be entrained by a sun they cannot see? And yet we know that a two-day-old adult bee that has never seen the sun shares the circadian rhythm of the rest of her colony, and if we separate her from her family the rhythm persists. Flying bees returning to the hive must somehow share the data of the sun's current movements in time with the rest of the colony to create a Hive Mean Time. But also, like mariners returning to Greenwich to resynchronize their ships' clocks for their next voyage, flying bees can individually recalibrate to the pooled metadata of all the other foragers' current experience of the sun in what is a collective colony clock.

The accuracy of the waggle dance is so important that if a forager is delayed for a significant length of time in the hive before she can perform her dance, without looking outside she can use her body clock to calculate how far the sun has moved since she last saw it, and factor this adjustment into the orientation of her dance to ensure that its data are still accurate at the moment she hits her dance floor.

The information the waggle dance communicates, like all social media posts, is optional, not compulsory: bees may prefer to trust their own memory of good sources or follow the scent trails released into the breeze by blossoms, but when food is scarce, food intelligence becomes vital, and sharing it via the waggle dance, essential.

Compared to most other pollinators, the mass super-pollination services provided by colonies of honey bees will actively make any flowers you plant for them more fertile, or stronger, or some combination of both decided by the plant. And it's all down to savings on the "free booze." Part of the energy bill a plant incurs when it reproduces is making the nectar it offers flying insects as an inducement to pollinate it. If a blossom is visited by no honey bees, and only one in ten more random pollinators visits carrying the correct species of pollen, nine drinks of "free" nectar haven't been paid for—the plant has produced them for no benefit. If the plant is visited by ten honey bees, all ten drinks of nectar are paid for in pollination services—at this rate the blossom might be fully pollinated ten times sooner, for a tenth of the cost in nectar. The plant super-pollinated by honey bees now has more energy-gathering time in the sun to reinvest these energy savings: it could decide to produce more blossoms to make more offspring—more food for your bees; or grow more energy-gathering leaves; or store more energy in its roots ready for next spring—or all of the above.

When honey bees super-pollinate so efficiently, all plants they visit share in the energy savings, even one being watered in a window box will benefit from the reinvestment of its energy. The interest rate on these savings may be modest, but it's compound.

A failing queen is the other main reason for a reduction in the pollen going into your hive. Nothing lives forever, and after years of reliable laying, every queen will start to become less productive. The bees quickly recognize any failure of the queen to take full advantage of the resources they're providing her, and in a process known as supersedure—familiar to politicians the world over—they raise a new virgin queen. They'll keep the old queen laying eggs until the virgin queen has mated and is ready to lay eggs too. Then the former monarch is dispatched. Usually all this is taken care of smoothly and discreetly with the minimum of fuss inside the hive, and all you'll notice is the sudden increase in pollen being flown in to accommodate the new queen's accelerated egg laying.

There isn't really anything for the idle beekeeper to do to facilitate or ameliorate this ruthless process. But we can help all our queens live long, healthy lives without resorting to the surgical opening of our womblike hives: instead of the surgeon's scalpel we can offer the physician's medicines—there's something we can grow that will help the bees throughout the year, with the added bonus of a delicious, healthy harvest for us too.

The forests in which the bees lived in tree cavities for fifteen million years would have looked very different from the ones we know today: humans now manage forests for timber, and the value of that timber is decimated by rot. So we quickly remove as much rotten wood from forests as possible to prevent it from infecting our investment—healthy trees—which we harvest when they're mature but not decrepit. Before our husbandry, forests would have had a much higher proportion of naturally decaying wood, lying around as an essential part of their ecosystem, returning carbon and nutrients to the soil.

Maverick mycologist Paul Stamets also keeps bees. He noticed that a raised bed of woodchips in which he was growing King Stropharia mushrooms was being visited by his bees—with extraordinary determination they had lifted relatively enormous chips of wood to expose the intricate strands of his mushrooms' mycelium growing below the surface. When he looked more closely he could see that the bees were sucking on little extracellular droplets that the mycelium had exuded like beads of sweat. And they couldn't get enough of this underground fungus juice.

Stamets discovered that these droplets contain acids, enzymes, and all sorts of messaging molecules: the mycelium is self-organizing and self-educating—as it grows and penetrates its environment, the tips of the mycelium explore anything new or unfamiliar, and then intelligently change themselves to find solutions to fight new toxins, or build new enzymes to digest new foods. And then these new solutions are shared throughout the fibers of the mycelial mat, which can extend unbroken for miles.

Messages can even be shared with other organisms: six bean plants were placed next to each other, their roots in separate pots; one plant was isolated from the other five and infested with aphids; the infested plant produced anti-aphid alkaloid chemicals to defend itself, the other five uninfested plants did not. The experiment was repeated but this time all six plants' roots were still isolated but growing in the same mycelium-rich soil; the isolated infested plant produced the anti-aphid alkaloids as before, but now so did all the other five plants that were not exposed to any aphids—the prompt for the plants to produce the aphid defense in the absence of aphids came from the mycelium.

We've known for some time that bees use fungus in their food preparation: "bee bread" is a staple—it's pollen fermented with wild yeasts and bacteria like we use to make sourdough bread from flour, or cheese from dairy where, in partnership with bacteria, fungi give us camembert, brie, Roquefort, and stilton. As well as preserving the pollen, the bacteria and fungi help to break down the cell walls of the pollen grains, and this process of predigestion releases the nutrients contained within to maximize their bioavailability to the bees.

But Stamets discovered that the mycelial droplets his bees were so keen on were specifically and spectacularly antiviral too. They upregulated the bees' immune systems. It turns out they are also significantly effective against human viruses such as H5N1 influenza and herpes.

P-Coumaric acid found in the droplets is used by both us and the bees to control our detoxification pathways, but this chemical is more essential to the bees—without it they can't detoxify, and it's not just pesticides—insect killers—that can accumulate to lethal levels in bees: herbicides like the ubiquitous glyphosate have now been shown to adversely affect bees' navigation and spatial learning after a single exposure to concentrations recommended for agricultural use.[5]

Although detoxification is important for all bees, it's absolutely vital for the queen: compared to worker bees whose lifespan in the summer is nine weeks, queens can live for many years. And during those periods when she's laying her own weight in eggs every day, she clearly needs to consume more than her own weight in food every day to give her the energy to perform the astonishing feat of converting so much plant protein into bee protein. So any toxins that tend to accumulate in the bodies of all bees will accumulate in the queen in spades—because she's consuming so much more food than any other bee, she'll also be consuming huge amounts of any toxin it contains and for far longer than any other bee.

This is not just bad for the queen. If any toxin damages bee DNA in any way, as sole egg-layer and singular custodian of the DNA of the entire colony, a queen whose detoxifying system is not optimal will concentrate the damage and may pass it on to her offspring.

P-Coumaric acid is just one of a multitude of medications available in mycelium that we know of. The bees may know of many more: mycelium is the pharmacy they have safely relied on for millions of years. We have simultaneously deprived them of it by managed deforestation and increased their toxic load with intensive agriculture. But

5   http://jeb.biologists.org/content/218/17/2799

you can become an apothecary and open a branch near your hive: it's easy to grow mycelium in a shady part of your garden or a tub. Your local tree surgeon will probably be delighted to let you have a load of fresh hardwood chips, and for less than the price of two London pints of beer you can get two pounds of sawdust inoculated with spores of King Stropharia—enough for a bed of chips a yard square and six inches deep. Kept moist, this bed will provide mycelium for the bees to mine for medication within months, and next spring you will begin to be able to harvest mushrooms that will keep emerging for up to three years, leaving your garden soil vastly enriched, or your tub full of the finest mushroom compost. You might even mix some with bee-friendly seeds, press firmly into a hand-sized fiber pot, and idly lob into any neglected public soil to invigorate your neighborhood with beauty, fertility, and bees.

King Stropharia mushrooms are huge, meaty, and best eaten young, cut thick and sautéed in butter. Delicious piled high on a slice of toasted sourdough. And while you're enjoying your share of the pharmaceutical feast outside the hive, consider that all you've done so far to help the bees inside has been achieved without attempting to shoot magical silver bullets: it's more like you've stitched silver linings to some of the clouds that can put their prosperity in the shade. But there's a creature that can make even the most pacifist beekeeper rush to the gun cabinet as it eats bees like you're eating mushrooms: the parasitic mite *Varroa destructor*.

Unknown outside Asia in the 1960s, in the past fifty years this mite has achieved world domination—it is now present in almost every beehive on the planet. It only managed to traverse the globe at such speed with our help—humans posted bees to other humans in other countries and the varroa mite hitched a ride. It feeds by drinking the blood of baby bees. Imagine a newborn human baby in an incubator with something dark brown, the size of a golf ball, crawling over its body and then sinking its teeth into the baby's skin where it not only sucks the baby's blood but infects it with diseases. For bees these include the prosaically lethal Deformed Wing Virus.

The instinctive desire to instantly exterminate this parasite is under-standable, but not so simple to achieve with chemical poisons—you're trying to kill a bug on a bug, and what's toxic for the mite is toxic for the bee it's attached to.

Accurate dosing of your poison becomes critical for the bees' survival of the treatment—the bees will always experience some harm in the process, the mites become resistant to the very latest miticides with astonishing speed, and because those lazy drones from other hives are always welcomed by your worker bees, hours after treatment your hive can be reinfested with healthy mites unwit-tingly carried in on the bodies of male visitors from infested hives you can't control.

If you'd succeeded in sticking around for fifteen million years negotiating all kinds of onslaughts, would you suddenly outsource your future survival? Let alone to a life-form that's only been around a few hundred thousand years and has already demonstrated its stag-gering fallibility by inadvertently spreading the parasite it's now try-ing to kill? That's not the kind of caretaker I'd bet the farm on.

Ultimately the bees will have to take responsibility for them-selves and learn to coexist with the varroa—hopefully by chucking them out of the hive at a rate that keeps their population below the numbers that can harm the bees. It appears that some colonies are already starting to exhibit this "hygienic behavior" and, in the face of the extraordinarily unnatural speed at which we introduced it, have learned very quickly that this mite poses a threat they need to deal with.

By housing your bees in an idle Warré hive, you've already stitched a silver lining to the dark cloud of varroa: for most of their life cycle the mites prefer a cooler temperature than the bees, so they will choose the colder corners of the hive boxes, as far away from the bees' warmth as possible. But because you've insulated your hive, those corners aren't so cold—they're cozier for the bees and now there's nowhere ideal for the mites to thrive: you're not evicting, but you are discouraging.

Varroa mites don't like high humidity—they do less well in the tropics. The walls of your idle Warré hive are warmer because you've insulated them and this means there's less likelihood of water vapor in the hive atmosphere condensing on cool walls and lowering the humidity. You can see this effect if you take a hot shower in winter in a bathroom with a single-glazed window: the condensed water running down the cold surface of the glass used to be in air in the room. Now it's not. The extra insulation of double-glazed windows not only lets you see the frozen winter view when you're showering, it keeps more moisture in the air you're breathing—its humidity is higher. Bees control the humidity in the hive and by insulating the walls we help them keep it at the level they prefer: higher than varroa like. This tips the balance against the mite and in favor of the bee.[6]

A recent Norwegian study[7] shows that colonies whose bees spend marginally less time pupating are much better at surviving with varroa: if their larvae spend 280 hours developing into an adult bee instead of 288, that's eight hours less for the mites to safely munch them inside their sealed honeycomb cell. This 2 percent reduction in time results in a 10 percent increase in adults hatching earlier in survivor colonies compared to colonies that varroa kills. That 10 percent appears to be a matter of life and death. But the bees that hatch after 280 hours aren't incomplete or partially formed. All the biological processes necessary to fully develop larvae into adults have taken place, just more quickly. And the simplest way to make all those synchronized biochemical reactions happen sooner would be for the bees to very slightly turn up the heat—if that's the case, your woolen wrapping will help.

6   Zachary Huang, "*Varroa* Mite Reproductive Biology," *Bee Culture* 140, no. 10 (October 2012): 22.

7   M.A.Y. Oddie, B. Dahle, and P. Neumann, "Reduced Postcapping Period in Honey Bees Surviving *Varroa destructor* by Means of Natural Selection," *Insects* 9, vol. 149 (2018).

None of the silver linings the insulation of your idle Warré hive adds is as instantly dramatic as the promise of the silver bullet of a toxic chemical treatment for varroa, but they are all persistently hostile to the mite and beneficial to the bee every hour of every day.

Until more bees begin to pupate faster or adopt hygienic behavior, there will be losses: even in the best insulated hives some colonies won't be able to adapt quickly enough and will be overcome by varroa. If you only have one hive, how do you quantify the risk when local beekeepers say they're experiencing 30 percent losses? 60 percent losses? If you lose your one colony, you cease to be a beekeeper. And in these circumstances I fully understand the impulse to treat with chemicals. I've never done this because I believe it selects for stronger survivor mites and bees that can't adapt, but I'd recommend treating any skull and crossbones symbols with the greatest respect. And the old saying "If all else fails, read the instructions" still applies to the newest of new concoctions.

But it might be more fun to use those indecipherable odds of losses to justify going plural—getting additional hives and more bees, spreading your risk, and shouldering the onerous burden of harvesting even more honey.

8

Bee Gentle

There is an old superstition that the bees must be told. From the ancient Egyptians to the Celts, many cultures have revered bees as messengers of the gods and bearers of wisdom from the spirit realm: a living connection with our ancestors. "Telling the bees" requires the bees to be verbally informed by the beekeeper of births, marriages, arrivals, departures, and deaths in the family. Before digital record-keeping, people traditionally used to write much of the same information in their family Bibles—that is, people who could afford a Bible and knew how to write in it.

But long before printed Bibles, paper, or even handwriting, humans have felt the need to somehow record their lives in a living culture that will survive them. From Celtic warriors who believed a good death in battle would immortalize them through unforgettable songs of their extraordinary skill and courage sung forever, to our forensic analysis of the tiniest fragments of DNA to identify the remains of particular individuals in disasters that have killed many, we need it to be known by the future that we came and went, and left a mark.

When a beekeeper died, it was the superstition that their bees not only had to be told, but they had to be involved in the mourning; otherwise they, too, might depart. Sometimes this became quite involved: when the beekeeper's coffin was lifted at their home to be carried to the church for burial, at the same moment their hives would be symbolically lifted up a few inches from their stands, which

OPPOSITE *"Over here!"*

were draped in funereal black. The church bell, tolling as the coffin approached the empty grave, would stop as the coffin was lowered into the ground, providing a useful, long-distance sound cue for the beehive to simultaneously be lowered back to its original position as the bees' keeper returned to the earth from which they came.

Long before wedding planners spurred on a multi-billion-dollar global industry, hives would have been lavishly garlanded in the bee-keeper's wedding decorations and a slice of wedding cake left out for the bees, presumably to encourage reciprocal fertility in the happy couple. Though I suspect that in a role-reversal of adults eating the mince pies left out for Santa Claus to provide children with evidence of his existence, here the bees would witness the little darlings steal-ing the cake.

You are of course very welcome to observe these ancient ritu-als in your beekeeping practice. I haven't yet, but describe them for two reasons: first, because they suggest that for a very long time we have trusted the bees as some kind of a repository of the things most important to us, and in doing so we have intimately connected the sur-vival of bees with our own—in the same way miners breathe the same air as canaries in coal mines, we feed our families from the same plants bees pollinate to feed their young.

Second, I'd like to put into some perspective what I *do* tell my bees, in words that I actually speak out loud every time I harvest honey from a hive in the autumn.

But before saying anything to them, we need to be sure that they can spare us some honey. "Hefting" is the art of estimating the weight of honey in a hive by lifting one side of it. An experienced bee-keeper will know by feel if the colony can spare the top box of honey. But a digital luggage scale and some elementary math can take some of the guesswork out of the calculation.

Hefting can be done at twilight when the bees have lost the nav-igational guidance of the sun and are tucked up in the hive for the night. The process shouldn't disturb them and at this time of day they probably won't disturb you.

Remove the heavy roof and the insulated quilt box underneath it—the bees will be confined in the hive by your top bar cloth. Flip up your insulation on the bottom hive box to expose its handles. Use your hive tool to *just* separate this box from the base—you're only trying to break the seal the bees have made between the two with propolis and you'll hear it crack with only a hair's breadth of movement, after which let the box down to its original position. You're aiming for the bees not knowing you're doing any of this—channel a cat burglar jimmying a bedroom window with a light sleeper inside. Once you've made sure the propolis seal is broken on all the sides of the bottom box, hook the luggage scale onto the handle on one side and slowly lift that side of the hive an eighth of an inch clear of the base and take a reading of the weight. Any more than an eighth of an inch risks bees coming out to see what you're up to. And just in case any of them decide to investigate the tiny gap that just appeared, make sure you lower the hive down onto the base slowly enough for legs, wings, and antennae to be moved out of the way—when bees are squished they not only die, they release alarm pheromones that get lots of other bees excited and defensive—ready to sting—not what we want at all.

Lift up the hive with the luggage scale by the handle on the other side in the same way and take another reading. While the bees are wondering if anything just happened, add the two readings together and you have the total weight of the hive. To find out the weight of the honey inside you need to subtract the weight of everything you've put in: all the boxes, top bars, and insulation. This will vary depending on the type of wood you've used and its thickness, so it's best to have weighed one of your boxes complete with bars before you've committed them all to the hive, but they tend to be around six pounds each, and the insulation adds a fluffy pound. You also need to subtract the weight of structures the bees have added, less than two pounds of empty wax comb per box, and of course just under four pounds of the bees themselves. The number you're left with is the weight of honey in the hive. Deduct about forty pounds—more than twice the weight of honey that bees in Northern Europe will consume to survive an average winter—your climate and their consumption may vary. What remains is safely surplus—all yours.

So if a four-box hive weighs a total of a hundred and twenty pounds, more than eighty pounds is honey and more than forty pounds is harvestable. If the top box is absolutely rammed with honey, it might contain as much as forty pounds, so you can safely remove it. In an average year you would hope to harvest the top box. In a year when both bees and food have been abundant you might be lucky enough to be able to take two boxes. Unless you are desperate for honey, it's best to only harvest whole boxes, leaving any fractions for the bees.

But if you are tantalized by the dilemma of one and a half boxes of available honey, you can repeat the weighing process but just heft the top box on its own and then heft the top two boxes. Calculation might reveal that your share of the honey is spread over the two boxes, not rammed into one and a half, and you can safely harvest them both.

Now that there is at least one hive box in your sights, we need to plan ahead for the moment we remove it from the hive: it's going to contain some worker bees toiling away, filling up the last few available cells with honey, and finishing capping full cells with wax to keep the honey fresh—it's never not work-in-progress. But all the bees, however dedicated to their specific job, need to know that all is well in the hive, and this assurance comes in the form of their queen's scent being constantly circulated by one and all. When you remove the box you're going to harvest and disconnect it from the hive, the warm current of air rising upward that carried the queen's pheromones from below stops abruptly. It's not just cold, it's lonely—the bees inside your harvest box will have suddenly lost their mom. Torn between doing their job and finding their queen, the bees need us to help them choose the latter and return to the main body of the hive. Abandoning the honey to us.

The "Madonna Escape" is my version of this persuasion to return to Mom, and if I can make one, so can Jean-Paul Gaultier, and so can you.

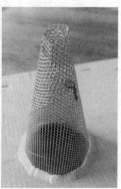

It's a lid that you place on top of the harvested box. Light pours in through four circular holes easily cut with scissors into a sheet of lightweight Correx—what real estate agents use for their For Sale signs instead of the lemongrass oil we use to lure bees to our bait hives—the darker the better to guide the bees to the bright holes that offer the only way out: because you've also made an identical wood and Correx structure without any holes to place as a bee-tight tray underneath the harvested box to contain dripping honey and prevent anything getting in.

The only way out is up—through cones made out of aluminum fly mesh that will guide the bees through an opening at the apex that is 0.35 inches in diameter. This will allow all the bees to leave, even the biggest, fattest drone that was holed up in the box, but prevent any of them coming back. It's a one-way bee valve and it's easy to make: instructions are in appendix 4.

You'll need something like a long cheese wire to ensure separation of the comb from the box below—I use an old top E string from a guitar tied around two handy pieces of wooden dowel, one at each end, but any thin wire or even strong fishing line will work well, and two short pencils or any old sticks would be as good for handles—make sure the wire's long enough to span the diagonal of the Warré hive box plus a good six inches for working.

Gather four small coins, a strap ready to bind your "Madonna Escape" in place, a water spray, and your smoker: you're now nearly ready for a dress rehearsal. Because harvesting your idle Warré hive is not only grand larceny, it's the most invasive thing you will ever do to your bees: like robbing a bank in broad daylight. And the most important thing we can do to mitigate the disturbance is to make it as brief and efficient as possible: a high-speed, precision heist that will limit the charge sheet to theft and avoid indictments for assault, murder, and infanticide.

When you remove the top box of honey, you're turning the living hive below into a chimney: all the lovely circulating ecosystem of heat, pheromones, and humidity loses its lid and instantly rises upward in a draft that draws cold air in through the entrance to replace the warm, moist, fragrant hive air that is rushing out the top—babies are chilled, every process in the hive is disrupted, and intruders like wasps or opportunistic robber bees from other colonies can invade the vast indefensible opening you've created. When this situation occurred during the millions of years bees lived in trees, it would likely be the result of the tree snapping at its weakest point: the cavity. As we imagined with the thin-walled tree cavity in the Natural History Museum, the colony's home would be instantly destroyed. Imagine you're a nurse tending to premature babies inside a neonatal ward when without warning some superior power suddenly removes the entire hospital roof.

When my grandfather served in France during the First World War, painting a red cross on the roof of his military hospital tent made it immune from attack while surgeons operated inside. This was no imaginary tent. He described how in the freezing winters, after making an incision in a casualty's abdomen he would immediately put down his scalpel and cup his hands tightly over the opening he'd made—not to stop anything getting into the wound, but to take advantage of the human heat that was escaping from it—if he didn't, his fingers would become too numb, and he'd lose the sensation of touch he needed to accurately feel and mend his patient's wounds. I suspect that if he'd ever kept bees, their living hives wouldn't have been open an unnecessary second.

As quickly as possible, we're going to create a new lid for the hive on the box below the one we remove, and the first step is to get the new top bar cloth into place on the new top box. Like a tarpaulin draped over that roofless hospital. The second it's there the hive stops being a chimney: air stops rushing out and the delicate balance of the atmosphere in the hive can be restored by the bees—so the quicker we can make this happen the better.

I used to just carefully position a new top bar cloth on the hive by hand, but after a particularly disastrous attempt at a harvest on a blustery day—imagine trying to tarpaulin that roofless hospital in a gale of wind—I now favor a more reliable applicator: a four-inch-deep quilt box filled with wool. Cut a rectangle of the same breathable roofing membrane that waterproofs the woolen insulation on the outside of the hive. The short sides of this rectangle should be the width of a hive box, and the long sides should overhang it by a couple of inches at both ends.

Place the quilt box on top of the new cloth and temporarily pin up the two overhanging ends with thumbtacks: the cloth can now be easily and accurately positioned with one hand and won't get blown away by an inopportune gust of wind.

Here's what to practice with a spare hive box, or any cardboard box roughly the right size: put it on a corner of your kitchen table, which for rehearsal purposes is standing in as the hive. Place the Madonna escape on top of its matching tray somewhere as near as possible to the "hive" and, crucially, make sure it is stable. If this place can be raised off the floor, that's great, because that will make things easier on your back, but be able to replicate this elevation at the real hive on the day. Lay the new top bar cloth applicator beside the tray.

If you've ever done Tai Chi, channel it now. Put the Madonna Escape on top of the box; lift the box off the table and put it down on the tray; pick up the top bar cloth applicator and put it on the kitchen table where the box was. If you're going to be wearing gloves when you harvest—I wear kitchen rubber gloves—wear them now: this is how it will feel. You can even put on your bee suit.

Try it again and consider the Special Forces mantra: "Slow is smooth. Smooth is fast." Ideally you don't have to move your feet. Once you've got the swing of it, let's go deeper: when you put the Madonna Escape on top of the box, make sure you do the "star turn" so bees don't get crushed—lower it down to make contact at 45 degrees, then gently rotate to line it up. Without blocking off all the light, the Madonna Escape helps to contain the bees at the top of the harvest box while you lift it.

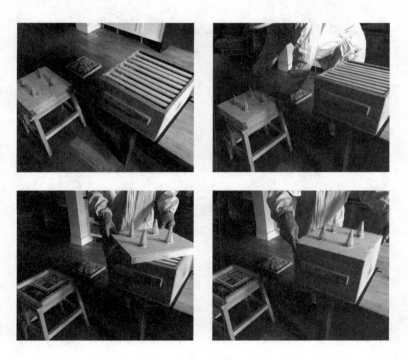

Aim for a similar star turn when you're putting the harvest box down on the tray.

But because the tray is very light and the harvest box may be dripping sticky honey, if you try the standard star turn, the tray will probably move as you rotate the heavy harvest box and you won't be able to smoothly align them. The way around this is to not rest the harvest box on the tray until it's aligned—in its eight-pointed star position, hold the harvest box floating an eighth of an inch above the tray and rotate it while it floats until you're aligned—bees will have been gently nudged out of the way, and the tray will not have moved. To misquote that legendary pugilist Mohammed Ali, "Float like a butterfly, don't get stung by a bee."

Now position one edge of the new top bar cloth applicator on the edge of the "hive" that's nearest to you, making sure the corners are aligned. A star turn isn't an option here because there will be honey that's dripped down onto the tops of the top bars you're aiming to cover with the new cloth—the moment the cloth touches this sticky honey it will become much more difficult to maneuver, and at the same time loads of bees may be coming up through those top bars to investigate the catastrophe. *Very* slowly, hinging at the edge touching the hive, lower the applicator into place: think of campers briefly trapped inside a softly collapsing tent, and gently encourage the bees down, back through the top bars, and into the hive as if only with the weight of the top bar cloth.

The applicator also tends to shield you from curious bees flying directly at you from the hive, and this may help to keep your movements calm and slow. You can even use a free hand to spray those bees down if they're alarming you, but try this out: multitasking might compromise your precision and now is a good time to find out—the worst that can happen is the kitchen table gets a little wet.

Practice all these elements together so your movements become confident, unhurried, purposeful, precise, and fluid. Graceful even. This is not so much to try and coin a new Tai Chi move, the "Honey Heist," but because this is the way of moving that your bees will find least threatening—like all of us they find sudden, unexpected movement alarming. It will also get the job done well, with the least risk of error and in the shortest time.

When you feel you have got close to combining sensei and sensibility, you are ready to harvest. A sunny day with little wind in autumn is ideal, and you want to plan to be lifting away your box of honey at high noon—when the sun is at its highest. These conditions will attract the maximum number of foragers to be working the blossoms outside the hive—a high density of arrivals and departures down at the entrance means fewer bees to be disturbed inside the hive as we work up at the top.

But we need cheese wire before Tai Chi. When we gracefully lift the harvest box off the hive, we need to know that comb filled with honey will not stick to the top bars in the box below, making our beautifully focused arcs of movement rip the comb apart, chunks of it falling to the ground, oozing honey. If you find this prospect disappointing, think how the bees will feel about it.

So a little before solar noon I put on my bee suit and gloves, but before I get to work with my old guitar string, like a witness in a courtroom I put my hand on the roof of the hive I'm going to harvest and I tell the bees. I tell them that I'm going to take some of their honey. That I'm only doing this because I know they can spare it. I apologize for the disruption I am about to cause, and undertake to keep it to a minimum. And I make them a promise: I will watch them carefully over the winter and if they start to struggle I will feed their honey back to them: not any honey, theirs.

I should confess that this is not the only time I talk to the bees. Every time I visit my apiary I rest my hand on each hive and greet its colony by name. The names I have given them celebrate how they arrived. I say a simple "Hello" to the bees because I want them to associate the vibrations in my voice with my smell and, crucially, the absence of harm when I'm there. When I tell them about the impending harvest, I hope that the same association of absence of harm might be communicated.

And for the beekeeper there is something about saying things out loud—even though you may have heard things like marriage vows spoken many times, intellectually understand and even agree with them, when you say them out loud in front of witnesses it's unforgettably different—standing at your beehive, more than forty thousand witnesses, however small, do something for your level of commitment to the promises you make.

Cheese out of the way, now we can go to the wire: remove the roof, the quilt box, and the insulation from the top box. The top of the hive is still sealed and the bees contained by its original top bar cloth. Stand at the back of the hive and nudge insulation down to expose the join between the top and second boxes. Work your wire into a corner near you at the bottom of the harvest box: slide it backward and forward until it's gone half an inch between the boxes. It's still not inside the hive yet, but it will allow you to push the blade of the hive tool in behind it, so you can prize the two boxes apart. They'll have been propolized together by the bees and you may have to work quite hard and wiggle around a bit before they crack open just enough for you to insert one of those small coins into the gap you've just made at the corner. The coin will hold the boxes apart enough for you to be able to make progress with the wire without any bees being able to escape.

Now work the hive tool into the other corner near you and prize the boxes apart so you can work the wire along and into the sides that join directly in front of you. Once that's done insert a small coin into the second corner, behind the wire. Now the harvest box is jacked up on the side nearest you by the height of the thickness of the two coins and you can start working the wire away from you, through the hive box. Slowly slide your wire from left to right, right to left, very gradually moving forward. Inside the hive the wire will be cutting any comb in the harvest box where it's attached to the top bars below. Sometimes this attachment is extensive, other times less so; you will feel the resistance to the wire as you move it. But even if there is very little to cut through, keep your wire's progress very slow—if you rush and accidentally slice through the queen you could kill the colony. I try and work the wire at a speed that I hope would be less annoying and disruptive to a queen than having to step over a neighboring bee that gets in her way: imagining how the mother of my children would have reacted had I, for some reason, tried to slide a skipping rope between her and the bed when she was changing position in the middle of labor.

As your ever-so-slowly-slicing wire gets nearer to the front of the hive, you may need to prize the boxes apart a little there and insert coins, but eventually you will get the wire all the way through and out the front of the hive and now nothing will be connecting the harvest box but gravity.

Carefully remove all the coins and let the bees settle for ten minutes or so, and light your bee smoker. I think of it like a fire extinguisher at a fireworks display—something you hope you'll never have to use, but it's good to have it ready and nearby. Enthusiastic smokers have all kinds of exotic favorite fuels, but I just recycle shredded documents as tinder and tightly rolled cardboard for a long burn duration—unforeseen "events" might make things take much longer than you've planned.

I only use smoke as a last resort. One of the bees' responses to smoke is to imbibe abdomen-filling quantities of honey, but to sting they need to be able to curl their abdomens down toward their victim: when it's full, such extreme bending of their abdomen becomes much more difficult—ask any heavily pregnant woman who's trying to tie her shoelaces—and though this increased difficulty may mean less stinging and lace-up pregnancy footwear, it doesn't necessarily engender greater calm. A spray of water mixed with a tiny amount of honey is a kinder alternative. The bees have a primal understanding of rain and how it interferes with their flight—it tells them "Go back into the hive and stay dry." Your sweetened spray will make for distractingly delicious rain as drenched bees drink themselves and each other dry, while you get on with taking a hive box full of their honey.

Place your Correx tray and new top bar cloth applicator in their rehearsed positions, and your Madonna Escape on top of the tray. Check that your water spray is working.

We're going to remove the old top bar cloth and deliberately, briefly flood the harvest box with daylight, because the queen bee sees it as danger and will shy away from it. If she happens to be unexpectedly up in the harvest box, she will scurry down to the safety of the lower boxes where it's darker. We do not want to harvest the queen.

The ticking clock starts now.

Standing at the back of the hive, peel back the top bar cloth from the edge nearest to you toward the hive entrance—this direction of travel ensures that the combs aren't orientated to shadow each other but will quickly let the light travel through the spaces between them and down to the bottom of the harvest box—no temporary dark corners for the queen to hide in while all that hive heat is escaping. The bees will have stuck the cloth to the edges of the top bars with propolis and this will provide some resistance, but don't worry, keep going and don't rush—you're wanting to give the queen time to leave the box if she's in there.

Workers will come out to investigate. You can stop peeling when you're halfway and spray them back down with a little water—a gentle misting, not a hosing!—but make sure you're keeping the cloth folded back so light can keep getting in: whatever she might have been doing if she was up in the harvest box, we don't want the queen thinking it's safe to return to it.

If the water spray isn't controlling the bees as much as you'd like and loads of bees come rushing out, use smoke: puff a few wisps of smoke across the top of the hive—never directly into it—and the bees will go down.

Thirty seconds elapsed.

Peel off the rest of the old top bar cloth. As you remove it there may be bees clinging to it—don't worry about them, just lay the old cloth on the ground near the hive entrance and they'll find their way in. Now maximum light is flooding down into the hive, but at the same time maximum colony heat is flowing uninhibited up out of the top box: more bees are coming out to investigate. Our agonizing dilemma is that we want to make sure we've allowed the light enough time to drive the queen down, but the entire colony is losing essential heat with every second that passes. I can usually stand it for about ten. And then I reach for the Madonna Escape.

At this point you know the drill and you and your body can become as one in the "Honey Heist"—with one caveat: even though I have harvested many times, the moment when I actually lift the harvest box off the hive I am always slightly surprised by how heavy it is. So although, like me, you may never be able to fully prepare yourself for this, we can at least make allowance for a little surprise and not be thrown by it. And when you're "floating like a butterfly" making that star turn with the harvest box an eighth of an inch above the tray, you will truly appreciate the irony of that phrase being uttered by the heavyweight champion of the world.

Beyond speed, accuracy, and efficiency, another advantage of your rehearsing is focus: you may find yourself less swayed by unexpected "events." Even though you're lifting the harvest box away from the entrance, a drop of honey might drip down there. Like a security guard in a bank looking out of the door and seeing dollar bills inexplicably floating down from the sky, guard bees may become extremely interested in your activity and rush out, flying up to persistently bounce off your visor, buzzing loudly in your face. The guard bees are sending a powerful signal for you to "Go away!" and it will work very effectively on you as an animal—you may irrationally doubt the protection your bee suit affords. But focusing on the familiar "Honey Heist" movements can help you override the very natural reflex fear the guards' behavior is precisely designed to engender.

In less than a minute, if all goes smoothly, you've separated and contained the harvest box inside the Madonna Escape and put an insulated lid back on the hive. This should be the single, most disruptive minute you inflict upon your bees all year—for the other 525,599 they may not be disturbed at all. But for now there will be lots of curious bees flying about, and this is a really good time to step back and let them settle down.

Activity around the harvest box will quickly reduce to bees gathering drips of honey that fell as you moved it. Similar activity will be going on at the top of the hive around the new top bar cloth. Once the bees have cleaned up and resumed normal duties, carefully unpin the new top bar cloth without disturbing its applicator, and pin it down to the box below. Nudge that box's woolen wrapping back up. You can now safely remove the applicator, replace your quilt box, and the hive will be back to full insulation.

For the harvest box the next few minutes are a little pregnant. What you are hoping to see is a bee: any bee, buzzing up inside one of your cones, climbing out of that hole at the top, and, maybe after a little hesitation as she realizes she can't get back in, flying off to find her queen back in the hive. Followed within ten minutes or so by growing numbers, and eventually all the bees in the box.

As the stragglers are flying in through the hive entrance, you can replace its roof, and strap the Madonna Escape to the harvested box so it can't be blown off by a gust of wind. At twilight you can complete this most enormous day of your beekeeping year by installing a mouse guard at the hive entrance. Mice can invade the hive to hibernate and steal honey, wreaking havoc in the process. There are various designs of mouse guards commercially available, but I just use a strip of metal with one small hole drilled in each end for thumbtacks to hold it in place so its bottom edge reduces the height of the entrance to three tenths of an inch. This is too narrow for a mouse to squeeze its head through, but wide enough for the reduced winter bee traffic to be unimpeded. I put a three tenths of an inch thickness of wood on the floor of the entrance and rest the strip on it as I fix it in place with the thumbtacks, and the hive is now ready to overwinter.

Unless . . . There are two rare scenarios that require that you beat a tactical retreat and put the harvest box back.

If no bees have emerged through your cones half an hour after you put on the Madonna Escape, it's highly likely that for whatever reason there are some babies in the harvest box with nurse bees that are refusing to abandon their young. Knock at the box as if it were a front door and listen carefully: if you hear a ripple of buzzing, it will be filled with bees. Lift the Madonna Escape and look through the top bars: if you can see bees between the combs, knock the box again and if there's a surge of buzzing but the bees stay put, the harvest box needs to go back on the hive before those babies get any colder.

You'll have to remove the new top bar cloth from the hive, lift the harvest box complete with Madonna Escape off the Correx tray, and return it to the hive with a floating star turn. Let the bees settle for five minutes, then replace the Madonna Escape with the new top bar cloth, and come back in two weeks and try again.

The other, even rarer, scenario is much more dramatic: if, when you separate the harvest box, almost straightaway bees start to stream out of the hive and cover your cones attempting to get into the harvest box, this is because their queen is inside. Urgently put the harvest box back! If the bees aren't settling when you've done this, check the Correx tray—the queen might have fallen from the comb during your maneuvers or crawled down there—if you see a ball of bees, she'll be in the middle of them for her protection. Take the Madonna Escape off the top of the hive and gently tip the ball of bees from the tray onto the top bars, covering them with the upturned tray as a temporary roof. Once the queen and the bees have settled, restore the hive. Come back tomorrow and try again: maybe this queen needs a little more light to encourage her to leave before you lift . . .

As your bees make a willing, collective departure from the harvest box through the handmade cones of your Madonna Escape, you will see increasing numbers of bees appearing at the hive entrance pointing their abdomens upward and outward as they fan out pheromones from their Nasonov glands to make a scent trail to guide the harvested workers the short distance back home.

There is something about seeing the bees help each other to voluntarily leave the honey behind that takes some of the sting out of my reservations about taking it from them. It feels like the colony has decided that caring for each other is the most important thing. As if the scent trail were saying, "Come home . . . ! Don't worry—it's only honey . . . We need you . . . ! All of you!"

9

Bee Productive

There was great excitement at the University of Cardiff when scientists discovered some Welsh honey that is as powerfully antimicrobial as the legendary Manuka honey from New Zealand, famed for its healing powers and exorbitant cost. The outstanding Welsh honey came from two hives kept by a druid in Tywyn, Cardigan Bay, but the scientists weren't interested in stone circles or solstices: the pollen of every plant species in Wales has been DNA barcoded, and the Cardiff academics were using this database to track minute amounts of pollen in the Tywyn honey, and thereby identify the particularly medicinal Welsh flowers the bees must have visited. There were even plans to carpet the roof of a shopping center in these flowers, providing Cardiff bees with the source for a commercialized Welsh super-honey (Cymruka™ . . . ?).

But the academics then discovered this isn't an entirely Welsh phenomenon: a hive near Southampton, England, also produced honey every bit as super as that from the two hives in Tywyn. And there's a connection between these three that also surpasses nationality: they're all unconventional hives that require a system of bee-keeping that aspires to be more bee-friendly. They're all Warré hives, and you've just harvested a box of honey from your idle one.

So what's behind the difference? And could your idle Warré hive intrinsically contribute to the antimicrobial potency of its residents' honey?

In conventional hives the nursery is separated from the larder by the queen excluder—worker bees have exclusive access to store honey at the top of the hive in a child-free zone that's also extremely convenient for the beekeeper but doesn't of itself deliver Manuka-like properties.

Nothing in your idle Warré hive reins in the freedom of movement of the queen. She can go where she likes, lay where she likes, and her entourage of worker bees will store honey and pollen to feed their young in whatever empty wax comb happens to be available nearby.

As in the bees' unfettered natural habitat, the older cells of the Warré honeycomb will have been flexibly reused for many different purposes after that initial building spree to accommodate babies when the swarm moved into the empty hive: pollen stash; nectar distillery; baby-gro; bee bread fermenter; baby-gro; honey store. And between each change of use, the bees will have scrupulously cleaned out the cells. But when you're converting a cell from raising babies to storing food, making a soiled diaper clean enough to use as a table napkin presents significant hygiene challenges . . .

Fortunately bees don't want their food or their honey to taste of baby poo either, and so after they've scraped out every last speck of detritus, they cover the surfaces of prospective honeycomb storage cells with the ubiquitous dark-brown disinfectant varnish of propolis. This is the same antiviral, antifungal, and antimicrobial concoction of enhanced plant resins the bees collect and use to coat every surface of the hive: its warm, humid confined space full of nectar and honey would be heaven for fungi, viruses, and microbes without it. Bees will use propolis liberally if they have to—the enormous body of an invading mouse that they've stung to death but can't physically remove will be mummified and entombed in a sarcophagus of propolis so its decomposition doesn't poison the hive. But flying to gather propolis consumes enormous amounts of energy, and efficiency rules everything bees do—they will always use the absolute minimum.

In a conventional hive, the cells used to store honey are never used for any other purpose. They have never harbored babies or baby poo. So we can suppose that a canny bee assessing how much propolis is needed to protect the flavor of its honey in this clean wax comb will think, "Not much."

Whereas an equally canny bee in your idle Warré hive planning to store its honey in cells of comb that might have contained *generations* of baby poo might think, "Quite a bit," and propolize lavishly to ensure that the honey will remain delicious.

Whisky drinkers will know that much of the distinctive flavor of their spirit is down to the water it's made with. Nowhere is this better illustrated than on the Island of Islay where the Ardbeg, Lagavulin, and Laphroaig distilleries nestle on the same coastline with only two miles separating the three of them. They're so close you suspect some of their sources of water must be flowing down different sides of the same hill. But the flavors of their single malt whiskies are uniquely distinct.

Similarly, beehives inches apart can be sourcing nectar from completely different sources of blossom, evidenced by the different colors of pollen their colonies are bringing in—they share food intelligence within colonies, not between them. The nectar from different plants will produce honey that tastes subtly different, but different sources of propolis from different plants have different chemical makeups and different flavors too.

The distilled spirit that goes into the whisky barrel is clear, much like vodka, but comes out years later a beautiful shade of gold. Because of what it absorbs from its contact with the inner surface of the barrel—the remnants of the sherry or bourbon that had previously been stored there, along with compounds from the wood.

Perhaps your idle Warré honey is significantly higher in antimicrobial compounds because it's absorbed them from the greater levels of antimicrobial propolis it comes into contact with inside the multi-repurposed cells of the honeycomb where it's stored?

Certainly the honey stored this way is what the bees themselves would naturally consume. And if on a distillery tour you were lucky enough to catch some of the workers helping the angels share out a crafty dram, wouldn't you want to taste the whisky from the barrel of their choice?

Before we open our tiny hexagonal barrels of gold, we need to remove the combs from the box we've just harvested. And this offers another opportunity for rehearsals, but this time we're going to learn how to use a tool—the Warré hive knife. And if that sounds scalpel-like it's because it is.

I'm sure the day will soon come when bee inspectors arrive with devices like the ones traffic police officers use to breathalyze drivers they suspect of drinking alcohol, or dogs like the ones that sniff your luggage at airports for drugs and paper money. Or even tiny microphones connected to smartphone apps that will analyze the health of the sounds your bees are making inside the hive like the ultrasound scans we perform on our developing babies from outside the womb. We may not yet be able to correlate the distinctive sounds bees make when they're suffering from specific notifiable, communicable diseases, but all those diseases have distinctive smells. Our noses may only be able to detect them at a stage when a diseased colony has already become an infection risk for others, but artificial noses will be able to detect molecular traces of disease even before signs are visible to the human eye: there would be no need to surgically open up healthy hives after detectors had sniffed their entrances and given them the all-clear. But for now bee inspectors are obliged to eyeball your combs. This means removing them from the hive when it's full of bees. And it's essential to practice using the tool that will enable you to do this, so you can minimize the harm the procedure will inevitably cause. Trainee surgeons practice their craft on cadavers, and your harvested box of honey provides a safe opportunity to practice going into the hive with a blade but without harm to life.

The Warré hive knife is a short blade at right angles to a long handle. Its function is to detach comb from the walls of the hive so it can be removed intact. You rest the long handle on the side of the hive box and lower the blade between two combs as if you were using it to spread butter on the surface of one of them. When you get to the bottom of the comb, you rotate the blade through 90 degrees so it's now under the side of the comb, and then you pull the handle slowly upward, resting the blade flat on the side of the hive box. When the blade reaches the top bar, you've detached that side of the comb from the box. Repeat the process on the other side.

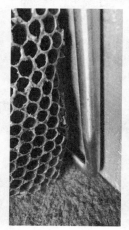

Knowing where your blade is helps enormously, so indicating the depth of a hive box with some tape, and the direction the blade is pointing by a line drawn on the handle, will let you know its position when you can't see it.

 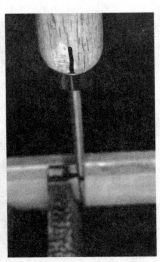

For an inspection you'd also need to use your wire to separate the combs from the box below, but as you've just done that for harvest, you can now carefully prize the top bar loose from its propolis seal with your hive tool and hold its ends to carefully lift the comb out of the hive. Practice putting it back exactly where it came from, and move on to extract the next comb. Imagine the hive box is full of bees you're trying not to crush. By the time you've done all eight in the box you're not only ready to be inspected, you've loosened the combs from which we're now going to release the honey.

Conventional wooden-framed honeycombs are spun in a mechanical centrifuge to extract the honey. The principle is simple and well understood but, like the spin cycle in your washing machine, difficult to make from scratch. And difficult to clean: imagine the inner workings of your washing machine coated in sticky goo after you've used it to spin-dry honey. For our freely formed idle Warré combs, we're going to blithely take advantage of something we profoundly don't understand but that is available everywhere without effort: gravity.

We will provide a sharp-pointed kitchen knife, a chopping board, some stainless steel mesh, and two large plastic food-grade buckets, both with one moving part—a lid—and our planet will extract the honey.

Turn one bucket upside down. Find something circular in your kitchen—a side plate is usually ideal—with a diameter a few inches smaller than the bucket's, place it on the center of the upturned base, and draw its outline. Put the lid on the other bucket and draw the outline of the side plate at its center.

Use the kitchen knife to carefully cut along the line of the circle you've drawn on the lid. Within the circle you've drawn on the base of the other bucket, use the kitchen knife to cut square holes the size of sugar cubes. They don't need to be exactly square, or even enthusiastically hexagonal—they're just holes for the honey to flow through—but enough bucket has to remain within your circle to provide structural integrity, so aim for a maximum of 25 holes, not 250. If the base of the bucket is particularly tough, you may need to use some form of post–Iron Age drill.

You've just made a honey extractor from scratch, and unless "wear and tear" is a personal vocation, it will outlast you.

Stack the bucket with the holes in its base on the bucket with the hole in its lid: all the holes should align and the buckets should be stable. Fit your sheet of mesh—as fine as a standard kitchen sieve—snugly over the holes at the bottom of the top bucket and position your dual bucket filter halfway under your kitchen table. On your kitchen table, position the chopping board so a little of it overhangs the edge above the buckets below, like a diving platform over a swimming pool.

We're now ready to Chop and Drain. Shut any open windows or doors unless you want to be joined by enthusiastic bees or wasps.

Roll up your sleeves, tie back your hair, wear an apron, and meet your tactile needs for the immediate future: make urgent phone calls, go to the bathroom, tune into your favorite radio station or podcast, be affectionate to your loved ones, because you are about to acquire a golden touch—and it's going to be sticky.

Lay a comb on the chopping board. If the beauty of its structure and the effort and care that went into its construction make you feel squeamish standing there with a big sharp knife at the sacrificial altar to your sweet teeth, console yourself with the thought that this comb has served its purpose. In the hive old comb is torn down and rebuilt by the bees—they will recycle everything and so will you. With the kitchen knife, cut the comb in half along the axis of the top bar. At right angles to the top bar, slice down through every cell at the edge of the bottom half, working cell by cell away from the buckets. Your knife is leaving a sweetly oozing swamp of honey and bits of comb in its wake, and before it seeps off the chopping board use the side of the blade to scrape it over the edge where it will fall with a deliciously wet thud into the bucket below.

Once the bottom half of the comb has been dispatched, carefully cut the top bar from what remains: you're going to reuse this top bar and by leaving half of the top layer of wax cells attached to it, you will be providing the best possible guide for the next bees that will start building new comb to hang from.

As you work your way through all eight combs from the hive box, you may encounter cells filled with pollen or other stuff you can't quite identify—it's all good—keep chopping and scrape it into the buckets. The only exception I make is for things that are moving: wax moths are nature's comb recyclers and are raring to go in the absence of bees or humans. Sometimes you will find their white inch-long larvae enthusiastically wriggling about as they munch away. They do not spoil the honey. I throw them outside to give the birds an unexpectedly sweet, protein- and fat-laden treat.

When the last chopped comb has dropped into the top bucket, put on its lid to keep everything else out. Checking that the honey is dripping through to the bottom bucket will be irresistible. Cleaning up your kitchen less so. But if the honey-coated chopping board can be spared for a rain-free day or two, take it along with the hive box and all the top bars and leave them a little distance from your hive[8]—grateful bees will take every last morsel of honey, and your newly non-stick hive box will make half of a seductive bait hive next spring—fully propolized, and with remnants of wax comb built onto its walls, it could not more authentically smell of desirable residence to a scout bee.

Back inside, what you do with the temptation to lick your kitchen clean is a private affair.

Depending on how warm it is, it may take a good few days for all the honey that's going to drain out of the chopped comb to make its way into the lower bucket. Lighter "impurities"—my favorite—will have floated to the surface and can be skimmed off and devoured.

---

8   Too close to your hive entrance and robber bees from other colonies, attracted by the free honey you're offering, might be tempted to go inside and steal more.

The honey beneath is the equivalent of wholemeal bread containing all the goodness of the hive, but if you prefer the clarity of more white-floured refinement, you can repeat the filtration process with finer mesh.

Use a clean jug to scoop and pour your honey into clean jars with tight-fitting lids to keep it dry. However finely you filter it, all honey contains minute quantities of natural yeasts that will start fermenting it once moisture absorbed from the air makes the honey's water content high enough. Most commercial honey is pasteurized to kill this yeast and extend the honey's life once it's been opened to the air. Unfortunately the heating involved kills or deactivates almost everything else. Enzymes, antibiotics, and countless as yet unknown beneficial compounds the bees add to the nectar as they distill and store it are the casualties that leave pasteurized honey the nutritional equivalent of sugar syrup. Your bees didn't add all these enhancements by accident—they put them there to help nourish them through the winter, and they can nourish you too. Your unheated honey is raw: enzymes and everything intact, the way the bees made it for themselves. Keep it dry so you don't have to share this golden nourishment with the yeasts.

Unless of course you positively want to: there's fun to be had deliberately adding a measured amount of water and allowing the honey to be transformed into a heady golden pleasure. By carefully rinsing the last traces of honey that are still sticking to all the bits of chopped-up comb in our upper bucket in spite of gravity's best efforts, we can add yeast to feast on the honey. And in return for so generously giving it something we can't have, the yeast will give us mead: the drink that has deliciously intoxicated cultured civilizations since Aristotle hit the bottle.

Wash your hands. Don't wash your lower bucket that you've just emptied of its honey, or the sticky jug. Dump all the sticky chopped comb into it, and stand the now empty upper bucket back on top. Fill the sticky jug with the clean water you're going to use for your mead and dribble it down the sides of the upper bucket. Your aim is to rinse as much of the honey clinging to it down into the lower bucket, but to use the least amount of water—less than a pint ideally. Feel free to use fingers.

Once you've got the upper bucket so clean a bee would ignore it, put it aside and mix your rinsings with the chopped comb in the bucket below. Handfuls of wet choppings are good for scouring honey off its sides.

At this point you need to make a choice that will reveal your priorities if not your character: mead over honey or honey over mead? If you want a gallon of mead, in addition to your rinsings you will probably need to add some honey from your jars to bring the concentration of sugars up to the levels that quantity of mead will require. If you only want as much mead as your rinsings by themselves will allow, you need to carefully add just enough water to the chopped comb to dilute the concentration of sugars down to the level your mead needs: no jars of precious honey will be added, but there will be less mead.

I like to make a dry sack mead. "Sack" means over 14 percent alcohol, and we achieve this by providing enough sugar for the yeast to make that much. We can measure the "enough" with a hydrometer. In the same way we float more easily in uplifting seawater than in a river because of the Archimedean weight of all the dissolved salt, a hydrometer will float higher in water that's had sugar or honey dissolved in it: how much higher tells us how much sugar for the yeast. Hydrometers are calibrated glass floats readily available from home brew suppliers. If you lower one into pure water, the reading at the surface will be 1.000. When there is sufficient honey in the water to raise that reading to 1.120, we have enough for sack.

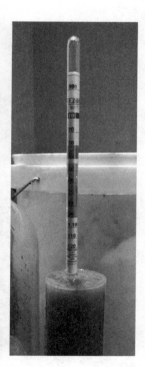

If you're not planning to use any jars of honey to make your mead, pour off a little of that pint you've mixed with the choppings and measure it with the hydrometer. It should read more than 1.120. Pour the solution back into the mix, swirl in a little more clean water, and measure again: how much your hydrometer reading has fallen will depend on many things—how much honey is stuck to how much comb, the sugar content of the honey, and how much water you've added. But it will give you an idea of how much water you should add in the next stage of this incremental process of lowering the reading little by little to 1.120, but not below.

If you want a gallon of mead, top up the choppings mix with a further seven pints of water and take a reading with the hydrometer: it will be lower than we want. Stir honey into the mix, measure, and repeat until 1.120 floats up.

Yeasts in the honey are already at work. But although they are wild and might be very interesting, like all of us they have a limit to the alcohol they can survive and it might not be as high as 14 percent. If they can only survive 12 percent alcohol, then they won't be able to feast on all the sugary honey we've provided and the mead will be sweet and less alcoholic. To ensure a dry sack I add some champagne yeast that can survive alcohol up to 18 percent and that will thrive until its food runs out.

Pour your yeasty rinsings into a glass demijohn, cork with a bubbler airlock, and wait for the "bloop" that will announce the coming fizz of fermentation. It will be weeks or even months before this blooping finally stops and you can rack off your mead and bottle it. Wait until there is a defined wet crust of pollen floating on the top and a clear layer of yeast sediment at the bottom before slowly lowering the tip of your unperturbing siphon tube into the mead and draining it into bottles. Clean, recycled clear glass wine bottles work well and reveal the gorgeous color.

Empty the freshly rinsed, chopped-up comb from your lower bucket into a large metal pan and fill it with water—if you live in a hard water area, use rainwater if you can. Heat and gently stir until nearly boiling.

Acquire the biggest, oldest, least-loved metal sieve you can find—just when it thought its usefulness was over, it's going to form a productive, if waxy, long-term relationship with you. Steal some of the cardboard egg trays any teenage children might be reusing to soundproof their bedrooms into recording studios, or find a less difficult source, less to hand.

Place the egg trays on the kitchen table where you chopped the honey and position the lower bucket on the floor beside them. With great care, pour the hot liquid from the pan through the sieve and into the lower bucket. Very quickly the sieve will fill up with solid gunge that chokes the flow. Stop pouring, and empty the contents of the sieve into the middle of an egg tray, banging it like a barista bangs out hot pucks of coffee grounds to empty the filter. Pour, bang, pour, bang, until all the hot, brown liquid is safely in the lower bucket and then put on its lid.

Spread the dark-brown solid gunge across the egg trays as soon as it's cooled enough to touch. Aim for a depth of half an egg—so you can *just* see some cardboard still defining each egg space. Place your egg trays somewhere warm and dry. They'll be wet with water from the gunge and they need to dry out as quickly as possible—there are still enough traces of honey in that gunge to feed mold if damp conditions persist. Once it is completely dry, you can tear off half-egg-sized portions to feed to your chickens or wild birds, who will devour all the nutritious bits of bees and old cocoons they contain. Or you can use them as firelighters for the charcoal in your barbeque—the traces of beeswax make them highly flammable and the dry cardboard is easily lit.

When you take the lid off the bucket next morning, it is as if night has become day—its murky, dark-brown surface will have been transformed into gold that seems to glow like the sun. Like fat rising to the surface of a cooling meat stock, pure beeswax has floated up and set hard into a circular disk, the soft rainwater ensuring its glossy shine. Lift up this waxen crowning glory and scrape and rinse the pollen that's stuck to its underside back into the brown liquid below.

There is nothing more we can usefully extract from these dark dregs, but they are excellent fertilizer, and if you feed them to flowering plants, next year your bees can recoup this investment of the last of what you've taken from their hive.

As with mead making, where there is a mellifluous world of variety and expertise that could infuse your every waking moment, so with beeswax candles. In England their production was controlled by the Worshipful Company of Wax Chandlers who were granted a Royal Charter in 1484 by King Richard the Third—the double prince-murderer wasn't all bad. But back then all candles were not created equal: twelve years earlier King Edward the Fourth had granted a Royal Charter to the Tallow Chandlers Company, which catered to very different ranks of society. Tallow is rendered beef and mutton fat, and when it burns it gives off smoke and an even fouler smell than the one that brutally informs you that you've lost track of time and the joint of beef or lamb cooking in your oven is now a ruined crisp. Tallow candles were cheap, used by the poor and for outdoor street lighting in the City of London.

Vegetarian beeswax is quite a different matter. If you didn't want your church to smell like a charred charnel house as its worshippers tried to get spiritual, the Worshipful Company of Wax Chandlers had just the light source for you. At a price: beeswax was eye-wateringly more expensive than tallow, so it was the preserve of the rich. Beeswax candles illuminate fragrantly and connect with the sacred practice of burning incense: frankincense and myrrh were right up there with gold on the present list at the birth of Jesus Christ, but in pre-Christian times incense was burned for its wide-ranging medicinal properties. Incenses are plant resins, and it was believed that different plants produced resins with specific healing properties that you could take into your body by inhaling the vapors given off when they were burned. The golden color of beeswax is caused by propolis—made from the plant resins bees also collect and enhance for medicinal purposes. So beeswax candles don't just smell nice, they're all-in-one incense burners delivering medicinal aromatherapy.

Purity was the unique selling point of the wax chandlers and they promoted it religiously. They demanded standards of quality that would show the candlemakers' due reverence for their Maker— punitive fines for shoddy work were levied in beeswax. So in language their Royal Patron Richard the Third might understand better than that Leicester car park from which he was recently exhumed five hundred years after his death, "Mea Culpa" for what follows.

To make idle candles, climb down from your throne but save the cardboard tubes of your toilet rolls. Buy some twisted and braided cotton candle wick whose thickness is suitable for the diameter of your toilet roll—about an inch and a half (square braid wick size 1). For each toilet roll you will need two wooden cocktail sticks and some glaziers putty.

With sharp scissors cut two tiny V grooves at polar opposites— north and south—of the tube, top and bottom. Just deep enough to accommodate cocktail sticks spanning the equators. Pierce the wick near one end with a cocktail stick and slide it into the middle of the stick. Rest this stick in the grooves at one end, run the wick through the cardboard tube, pull it taut, then pierce the wick at the other end with the other cocktail stick in a position where it will be held in the grooves at its end.

You should now have cocktail sticks held in place at both ends by the taut wick, which you can adjust till it runs through the middle of the tube. Stand the tube up on a plate and seal it in place with a generous ring of glaziers putty pressed around its base. Wash and dry a used, large food can. Squeeze its top to create a pouring point. Break off pieces of your golden disk of wax, put them in the can, and stand it in a pan of boiling water till the wax melts. If the liquid wax looks particularly full of bits of impurities, pour it into another clean can through a nylon tea strainer and then back again. Pour a little into your cardboard tube and check for leaks in the putty—if all's well, fill up to the height of candle you want and leave it to cool. Once the wax is hard, tear off the cardboard, remove the cocktail sticks, and trim the wick.

You have turned everything you've taken from the hive to use. It's what your bees would have done and it's a measure of your respect for them. It will change how you interact with them, however subtly, and they may sense that. Your mead may taste mediocre, and your candles may look more like they've been made by a bungler than a chandler, but they will lighten your darkness as the nights draw in and the bees hunker down for the winter.

10

Close to Home

The truism that if ten beekeepers engage in a debate you will get twenty-seven different opinions would definitely provoke a collective buzz of disagreement—over those very statistics just for starters. But the one thing most of them would not dispute is that all beekeeping is local. The local environment in which bee colonies find themselves influences their behavior, and it is that behavior that beekeepers are attempting to accommodate. Your neighborhood defines you as a beekeeper.

The idle Warré hive particularly tries to accommodate the bees' need to provide a stable temperature for their young whatever the local weather outside their home—and it does this by cladding the hive with insulation.

When I walk five hundred yards west down the road I live on I come to a building so tall that at certain times of the year it casts its evening shadow over the hives on my roof. This building was recently clad with insulation too. On the night of June 14, 2017, that insulation caught fire, creating an inferno that engulfed Grenfell Tower, a residential block of 129 flats. Two hundred and twenty-three people escaped, but that night seventy-one people died horrifically, mostly from smoke inhalation.

It used to be thought that the bees recognized smoke as a primal symptom of an approaching forest fire and that they would retreat into the hive to quickly consume honey in preparation for the colony moving to a safer home. This allowed beekeepers to believe that smoke "calmed bees down." We now know from studying bees' preparations to swarm that an organized relocation of up to sixty thousand individuals depends on intricate, sophisticated, and widespread communication that relies on pheromonal scent signals as well as vibration: scents that smoke utterly overpowers and confuses, leaving bees unable to communicate to collectively orchestrate such a complex endeavor so quickly.

Current thinking is that smoke tells them the forest fire is near and their only chance of collective survival is to cluster in the deepest recesses of their cavity and hope the tree will insulate them from the inferno until it passes. They probably aren't gorging on honey to fuel sudden flight but filling their crops with as much of it as they can because it will increase their mass: making them take longer to be heated up to a lethal temperature, and thereby increasing their chances of surviving the passing fire.[9] Smoke sends a simple message that every bee can understand without sharing: "Get inside! It's the safest place!"

The night the residents of Grenfell Tower smelled smoke in their environment, they raised the alarm and summoned firemen who gave them the instructions to stay in their flats until the fire was extinguished. These instructions were based on carefully calculated assumptions about how quickly fire would spread in this concrete tower, set against how quickly firemen would be able to put it out. Because of these assumptions, as with the bees, mass evacuation was not part of the plan in the event of fire, and there was only one narrow staircase that served the whole tower.

But the building had been changed since those calculations were made. Their timings altered. The insulating materials newly fixed to the outside of the building were more flammable than had been imagined, causing the fire to spread more quickly than the firemen could control. With catastrophic consequences.

9   G. Tribe, J. Tautz, K. Sternberg, and J. Cullinan, "Firewalls in Bee Nests—Survival Value of Propolis Walls of Wild Cape Honeybee (*Apis mellifera capensis*)," *Naturwissenschaften* 104 (April 2017): 29, doi:10.1007/s00114-017-1449-5.

Flames were still licking out of the charred husk of the tower early in the morning when we joined what was to become thousands of local people bringing food and bedding and clothing to neighboring centers of community where bereaved and homeless survivors were being offered safe haven. Even giving favorite toys our children had long since outgrown but we'd secretly kept in case of grandchildren did nothing to counter the brutal truth that this was all too late.

Grenfell Tower was an example of social housing—basic but decent, affordable accommodation provided by local councils in the United Kingdom since the end of the First World War when the Housing Act of 1919 made housing a national responsibility. These "Homes Fit for Heroes" provided cheap, rented accommodation for working people who'd joined up to fight for their country and in only twenty years would be required to do so again. The motives behind the provision weren't entirely altruistic: existing poor-quality housing had yielded physically poor recruits for the armed forces and that was not in the interest of any nation seeking military advantage.

The new council houses replaced disease-prone slums and offered to everyone what had previously been considered luxuries: hot and cold running water, inside toilets, and electricity. Even my family of surgeons agree that more lives have been saved by clean water and sanitation than by all the scalpels ever sharpened. These homes were purpose-built for the well-being of their occupants, not just as potential soldiers, but to generate the peacetime bonus of increased productivity: healthier workforce, healthier economy.

Our idle Warré hives are purpose-built too. But not developed so they can be better flung by trebuchets into our enemies' camps as was the fate of medieval military bees. We haven't designed improved "clusterhives" to proactively disperse on impact with the ground in order to sting a wider target zone of people more effectively. Instead our purpose is to enhance the accommodation we provide for the bees to specifically facilitate their health and well-being in order to increase their peaceful productivity.

And to do this we've looked to emulate the tree cavities they've successfully occupied for fifteen million years. But in nature, the fungus that penetrated a tree's defenses to feast on its wood had no plans for bees: it was just hungry and munched away, oblivious to their existence. A woodpecker, tapping the trunk of that tree, listening to the subtle changes in the reverberations of its percussion, was purposefully discovering the extent to which that fungus had spread, because wood softened by rot is easier to peck out for a home big enough for its young—bees inheriting the vacated cavity it had enlarged were no part of the woodpecker's thinking. For all the organisms that contributed to making the tree cavity bees eventually chose to live in, the work they invested was its own reward, and they opportunistically benefited from those that preceded them. And without obligation: bees don't go back to woodpeckers with a snagging list of all the features of a cavity they consider less than ideal in the expectation of more, better pecking of wood. And woodpeckers don't return to their old nest sites expecting to be willingly fed an agreed quota of delicious baby bees in payment for their part in the construction of those babies' accommodation. If their food is being vigorously defended by its parents, woodpeckers will eat easier meals elsewhere. And if a cavity is not right for them, then bees will move on and look for one that suits their purpose better.

Scouts are the oldest bees with the longest experience of what the colony does and what will best meet its needs. Dozens of them will be searching for a new home as their colony makes preparations to swarm. Once a defendable space has been evaluated as a possible candidate, the scout will fly back to the hive and dance a waggle on the comb that speaks the same body language as the foragers' description of a food source, only this scout's dance announces her enthusiasm for the particulars of what might be their new home—dimensions, entrance size, orientation and height above ground, and location, location, location. All very familiar to real estate agents compiling the particulars of a property they're selling, but the big difference here is that scout bees are never "economical with the truth."

Ever. Why would they lie? They are not on commission: they have no other purpose than the common good. They dance the unvarnished facts. And as with the floral waggle dance, the better the scout considers a site for their new home, the more vigorously and enthusiastically she dances—she's literally trying to create a buzz for the candidate she's describing.

But even scouts that can dance like Salome cannot seduce easy support from their social network—though we can ensure that enthusiasm goes viral across the planet with a mere click of our fingers on social media, other scout bees can only endorse a candidate if they actually go and check it out. They will follow the original scout's dance for directions, fly to the cavity, and see for themselves. They will carry out the same rigorous full survey and report back their enthusiasm for the candidate. Candidly. And if this dance generates further enthusiasm, it does so without botnets, celebrity following, commercial algorithmic bias, or state-sponsored cyber manipulation hacking the bees' direct peer-to-peer trusted network.

A very successful financial asset management company recently advertised its unique selling proposition with a poster that simply showed a pair of well-worn business shoes. The company proclaimed that it never invested in any venture its representatives hadn't personally visited. To see for themselves, to scrutinize and interrogate, face-to-face.

A prudent financial company would never put all its eggs in one basket: it would spread its risk across diverse sources of income. And foraging bees similarly benefit from many waggle dances advocating many different and varied sources of food—the wider the choice, the better the nutrition. But the scout bees are not planning to guide their swarm into multiple new homes. All their choices have to be narrowed down to one: the best.

So scouts are sensitive to the vibe of newly discovered cavities hitting the dance floor. And if on visiting a potential new home they rate it better than the one they were championing, they'll change their dance to this new, superior cavity and even more enthusiastically strut its stuff to all the other scouts. Support for preferred candidates only increases on personally verified merit, with the buzz about less suitable sites fading away until the field narrows to two contenders being vigorously promoted. At this late stage the swarm may now be hanging in a cluster from a branch outside the hive with time and food running out, desperate for a decision to be made. The wax comb dance floor left behind, the scouts can now be seen dancing on the surface of the cluster, their vibrations directly rippling through the interconnected bodies of the bees that are clinging to each other to protect their queen. Even now it's possible that a new, even better candidate might emerge and build a following that could topple one of the final contenders, but the debate becomes heated: scout bees resort to very unparliamentary head-butting of rival dancers to try and put them off their step and make them see sense or shut up.

Just before a swarm moves into your bait hive, you may notice a huge increase in scouts hovering at its entrance, feverishly investigating. These are the last floating voters in the democratic choice, and when the swarm begins to move in, you know your bait hive swung them at the last minute. The scouts have roused the cluster into flight and will guide all the bees to fly directly to the winning candidate. Every single one.

The informed decision to choose their new home is quite literally thrashed out until it becomes quorate. And Professor Tom Seeley has calculated that this open, honest, and meritocratic honey bee democracy guides them to the best choice well over 90 percent of the time. If a financial management company could match this level of success, its clients would be as rich as Croesus.

But success in terms of human democracy is harder to define or measure, and if all beekeeping is local, all democracy is local too: at the base of every pyramid of democratic power are local people choosing local representatives. A hundred yards west down my road there is a beautiful, publicly funded building that was purpose-built as a public lending library in 1891 by civic-minded Victorians. From my roof I can see its elegant cupola with a small golden orb at the center of a weather vane. For over 120 years it's told local residents which way the wind is blowing, without too much spin.

Every few years space is made between the bookshelves for a pop-up: simple wooden booths are erected, each with a pencil attached by a piece of string. It's my local polling station, and thousands of local residents turn up to make their mark on a ballot paper. It's where we cast our votes in national and local elections.

Recently our local council decided they wanted to lease the library building to a private school. I joined a group of residents who got together to oppose planning permission for the change of use. Local democracy in action.

We were an incredibly diverse bunch, all with different, often very personal reasons for wanting the library to stay, and this issue seemed to transcend some of the extremes of local inequality. Four hundred yards west of the library is Portland Road: at one end, near the foot of Grenfell Tower there is a council block of tiny flats with walls and floors so thin the residents unwillingly know everything about their neighbors' nocturnal urinations save the color; and at the other end of this mile-long road, nearer Kensington Palace and international embassies, privately owned houses change hands for more than twelve million dollars.

I was particularly struck by one man in our group named Ed who combined extraordinary passion for his well-informed views with a fastidious desire for everyone else to be heard—however shy, however inarticulate, Ed would encourage and respect their views. A rare combination.

Our campaign ran its democratic course and at the final planning appeal the council voted to award itself planning permission to do what it wanted and lease the Victorian library building to the private school. We lost.

But before the building was repurposed, Grenfell Tower burned, and the council placed a moratorium on all its developments. The library building was reprieved.

When our group met we looked Ed in the eye and told him that we needed him to know we were crystal clear that the library had only been allowed to stay because seventy-one people lost their lives. Ed was now homeless. He had lived on the sixteenth floor of Grenfell Tower and escaped from the fire. For years Ed had been part of the Grenfell Action Group that had been raising the alarm about safety in the Tower. So many of their concerns fell on deaf ears that seven months before the fire he wrote in exasperation on their website:

> "It is a truly terrifying thought but the Grenfell Action Group firmly believe that only a catastrophic event will expose the ineptitude and incompetence of our landlord . . . . Unfortunately, the Grenfell Action Group have reached the conclusion that only an incident that results in serious loss of life of . . . residents will allow the external scrutiny to occur that will shine a light on the practices that characterize the malign governance of this non-functioning organisation."

What would the bees that democratically chose to live in my hives make of this local human democracy?

Honey bees understand job specialization: they all perform various, clearly defined roles in the hive at any given time. They know that they can be better nurses if they're not having to simultaneously build wax comb, that they can't clean the hive and gather pollen from flowers at the same time. So even though they don't build the cavities they live in, they might understand that the people who built and insulated Grenfell Tower were specialists, like comb builders, but on the outside. They might understand that among those builders there were many different subspecializations: architects, crane operators, concrete pourers, fabricators, plasterers, joiners, plumbers, and electricians who all had to work independently, but together. They might recognize that the complex schedules of work needed to employ fuzzy logic to negotiate unforeseen circumstances. And when work on the Tower was going well, they might have even thought of it as a hive of activity.

But they would be surprised that so few of these builders actually lived in the Tower they'd built.

Bees delegate the selection of a new home to specialists—most of them are too busy with their full-time day jobs back at the still bustling, nonstop hive to be able to spare the time and energy flying around house hunting. And they haven't yet worked in a sufficient variety of those jobs to be as qualified as the scout bees to make judgments about which cavities would be considered good enough to even make it through to the selection process. They utterly rely on the honesty and rigor of all the scout bees to ensure that their new home is at the very least protected from the elements and defendable: when a swarm follows its scout bees toward the home of its choice, there is no element of hesitation in this buzzing mass of broiling conviction flying with determined velocity.

They might understand that provision for the safety of both the construction workers and the subsequent occupants of the Tower was delegated to specialists who were experts in that field. But they would almost certainly be amazed that these specialists could learn important, relevant lessons about safety without witnessing catastrophic failures with their own eyes. That these specialists could somehow know about cautionary events only other people had experienced—people they'd never met, even people no longer alive—by simply staring at flat white things with markings.

Our illiterate bees would also be surprised that so few of these safety specialists actually lived in the Tower they'd helped build. Or any tower.

Even highly literate, well-educated, civilized people have been known to resort to head-butting in the heat of an argument. But it's rarely because they don't care. When two scout bees are engaged in that endgame of democratic persuasion, they're banging heads because both are now so absolutely convinced by their levels of support that their candidate for a new home is in the very best interests of the colony that they will physically attack anyone challenging this. Both scouts take their responsibility to the colony with the utmost seriousness, and a few bruises. They are not only going to have to live with the consequences of this decision: they are going to have to live *in* the consequences of this decision with those on whose behalf they made it.

Scout bees might find it odd that democratically elected officials who have accepted responsibility for housing the people living in Grenfell Tower didn't live with them in the Tower. Or any tower. But delegated the day-to-day running of the Tower to tiers of intermediary people so elected officials were even farther removed from the experience of those who did live there.

No self-respecting scout bees would take the Grenfell Action Group's word for anything. But they would have rigorously investigated and assessed its concerns to get to the truth of them, rejecting spurious irrelevance, but urgently presenting crucial facts. And when they danced their dance they would expect to be heard.

It would take less than five minutes for a scout bee to fly the mile and a half from Grenfell Tower to the local council offices in the Town Hall. If she could meet the leader of the Council for a bit of a chin waggle, on behalf of the bees she might make this case to her fellow representative of democracy: that all of us need to stay better connected. She would highly recommend listening to the lived and living experience of those central to a decision that affects their future. The people on the ground.

Our failure to heed local knowledge and expertise has caused untold havoc for man and bee.

A beekeeper visiting the hives of Warwick E. Kerr near São Paulo, Brazil, in 1957 saw worker bees slightly struggling to crawl through queen excluders that were unusually covering the hive entrances and slowing down the flow of traffic. The visiting beekeeper knew what queen excluders were for and where they should be—between hive boxes—so, with the best of intentions, he decided to correct the situation and help the bees a little by removing the obstructive excluders, but without talking to their owner first. Kerr had carefully placed the excluders over his hive entrances to prevent the queens and drones from flying out to swarm or mate with local bees: inside those hives he was developing a strain of bee better suited to the hotter climate in subtropical Brazil. His visitor accidentally released twenty-six swarms of African bees imported from a hostile environment with many more natural predators than European honey bees have to deal with. The African bees had evolved to respond to this increased threat socially: as individuals they're very similar to European honey bees, but whereas a maximum of only 10 percent of European bees in a hive might respond to a perceived attack, *all* the African bees in a hive might fly out to deal with a potential intruder en masse, chasing it up to a quarter of a mile, stinging all the way. Kerr's African bees escaped and cross-bred with local European strains to produce the aggressive "killer bees" that have spread into the United States and killed more than a thousand people. Now one of the most successful invasive species on the planet, "killer bees" were not an outcome wished for by Kerr or his visitor, and one that a minute's conversation might have avoided.

At the same time as one man tried to help some bees in Brazil, Chairman Mao was coordinating millions of Chinese people in that agricultural revolution of the "Great Leap Forward," killing millions of sparrows and then billions of bees. But it turned out that the birds and the bees were all innocent of causing the grain "shortages" that were blamed on them.

To drag Chinese agriculture out of the Middle Ages, Mao had decreed farming by communes controlled by the state. To improve food production he imposed radical innovations to cultivation methods based on the untested theories of Russian agronomists Trofim Lysenko and Terentiy Maltsev. Among the "improvements" was the practice of deep plowing: instead of turning the soil to the conventional depth of six to eight inches, farmers now had to dig down a backbreaking three to six feet. The mistaken belief was that this would encourage deep root growth, but in fact it just buried the bio-logically active nourishing topsoil six feet under—beyond the reach of the crops, which had to make do with the relatively dead growth medium of rocks and subsoil brought to the surface.

Another "improvement" was close cropping, whereby grain would be sown more densely in the belief that it would crowd out weeds and thereby increase yields because plants of the same class would not compete with each other. Although this rationale might have been ideologically in tune with a communist revolution, it had no basis in reality: even plants of the same species compete for resources, much like unreconstructed humans. Farmers who for centuries had been intimately acquainted with hunger knew that increasing the density of seeds sown beyond the level that soil nutri-ents could sustain would just cause more seedlings to die from com-petition—and their families could have eaten those seeds. They were forced to throw good food on the land and watch it die and become inedible. Revolutionary change will always meet resistance, but the baby of the farmers' expertise was being thrown out with the old bathwater of inefficiency.

The peasant farmers actually doing the work in the fields could see that crop yields were going down, but their local leaders were part of a political structure that dared not admit that Chairman Mao's improvements were detrimental, and they undertook to deliver the predicted bumper harvests of grain to the cities where it fed the urban population and was sold overseas for valuable for-eign currency. But this meant that there wasn't enough food for the people who had actually grown it: farmers were allowed to starve to death in the shadow of full silos of grain they were not allowed to eat in order to maintain the illusion of abundance. When they tried to tell truth to power they were denounced as anti-communist and silenced. The sparrows and locusts took the blame, but billions of bees were collateral casualties and at least twenty million people lost their lives because political leaders who had ignored them were not prepared to lose face.

It's possible that if Chairman Mao had gone out into the fields and had an hour's frank, open, and honest conversation with some of the farmers who worked on them before making the "Great Leap Forward," millions of lives would have been spared.

I suspect the honest scout bees living on my roof would approve of those financial asset management company representatives who wore out shoe leather visiting every potential investment personally. They understood that not everything is as simple as it first appears on a spreadsheet: in our experiment with the cashmere clusters that showed the headline 33 percent energy saving in our insulated idle Warré hives, the assumption could easily be made on a calculator that this energy saving would equate to a 33 percent increase in honey production. A closer look inside the hive tells a different, more nuanced story.

But maybe the well-worn shoes we should be wearing are not the business ones pictured on that poster, but the ones that belong to the people we're visiting. Let's stand in their shoes. Connecting isn't just about empirical facts, it's about empathy. We could try walking the mile of Portland Road in the shoes of the people who live at the other end.

The stall holder on Portobello Road from whom I bought the $5 moth-eaten pure cashmere sweater for the cluster experiment is named Paul Smith. There is another Paul Smith, also in the rag trade, who is an internationally famous designer with a business empire worth multimillions. A grand Victorian Notting Hill mansion off Portobello Road is home to the Paul Smith Bespoke Service. I doubt he gets many special requests for moths. The two men know of each other and when they pass by they greet with a respectful, and utterly egalitarian, "Mr. Smith" . . . "Mr. Smith." I like to believe that the twinkles in their eyes at these encounters might be mutual acknowledgment of both the disparity of their wealth and their common humanity—that they are fellow travelers on Portobello Road where, but for the grace of a capricious market . . . In such a moment they would experience what for bees is a way of life.

It's unlikely you will have chosen to wrap your idle Warré hive in anything either of the Mr. Smiths sells, but even the most basic wool from the scrawniest sheep has an advantage over the high-tech cladding used to insulate Grenfell Tower: it is fire resistant. Everything burns—consider the stars whose ash we are—but wool makes it very difficult for fire to spread. Not only is its ignition temperature high—more than 1000 degrees Fahrenheit—wool needs more oxygen than is available in the surrounding air to allow it to burn. Its fibers have a cross-linked cell membrane structure, and when wool is heated anywhere near its ignition temperature they swell up and form an insulating barrier that stops any flame spreading. So if some powerful heat source does actually manage to ignite it, wool will briefly smolder before self-extinguishing without melting or giving off much smoke or toxic gas.

All this from applying sheep to grass.

In the green and pleasant United Kingdom, about 60 percent of agricultural land is of such poor soil quality that grass is the only thing that will productively grow. This pasture can't feed us wheat or potatoes, but it nourishes grazing livestock, and on its steepest slopes sheep grow wool from the scrag end of the chlorophyll. The main problem with wool as insulation is that it's delicious. Applying our collective wisdom to the search for chemical-free ways to deter moths and other insects from gorging on it would benefit everyone, with the possible exception of one Mr. Smith on Portobello Road. I like the angle of the research that's trying to find out why moths won't eat nutritious spider's silk, and somehow applying those natural arachnid properties to wool fibers. Maybe we'll find the solution on the Web.

Solutions to promote human interconnection have been peddled in watering holes across the planet since the dawn of commerce. Before the ban on smoking in public places, the carpets in pubs in the United Kingdom had to be made of pure wool. No man-made fibers were allowed. Not out of respect to customers who were deep-pile connoisseurs but merely to comply with fire regulations—in a place where people are by definition not entirely sober, burning cigarettes landing on flammable floor coverings were an avoidable catastrophe where the carpet was the easier of the two protagonists to control.

Landlords may have been able to save on their pub's carpet budget by increasing the percentage of polyester, but that hasn't been enough to stop thousands of them going out of business since the smoking ban. The pub across the road from our library is one of them. It's closed down. And as we lose these bowers of boozy bliss where rich and poor, all and sundry could meet friends, greet strangers, rub shoulders, drown sorrows, share laughter, sink pints, and float ideas, libraries are becoming some of the few remaining community hubs left to us.

Spurred by the circumstances of its reprieve, our local group is working to ensure that our library is adapting to meet the changing needs of its community, combining the virtues of the virtual with the appeal of the real. As it expands and transforms the range of its cultural offerings, those framed instructions of "Silence!" that used to adorn otherwise bare library walls are no longer so absolute: local artists' work is now quietly displayed like hanging gardens of the imagination beside screens of pixels that procure the planet. And time is made for music and the spoken word. Performed live. Face-to-face.

And books too. Like foraging bees, books still fly freely out of this hive of information to local homes where they silently fertilize minds with knowledge and culture, before returning to the lending library to await the interest paid them by their next borrower. Like the social interconnections the bees make with each other, human interconnections don't have to be fully witnessed or fully understood to be respected and valued. There may be a copy of what's in front of you on one of those shelves in that saved library building. Or out in any of the neighboring homes. In the hands of someone just like you. Reading it. Right now.

On the longest night of the year, the winter solstice, we gather friends. I save the previous summer's now one-year-old mead for the occasion. We draw the curtains and switch off all electric lighting: it's dark as inside a hive when we light home-made beeswax candles, fill everyone's glasses with mead, and propose a toast. On the night of December 21, 2017, it was this:

"Ladies and gentlemen . . . Directly above our heads—twenty-five feet up, on the roof—are thirty thousand bees living in six hives. Right now they're very much like us—on the longest night of the year, they've gathered together in a huddle for warmth and kinship. But six months ago, at the height of summer, they were busily harvesting sunlight in the form of nectar they collected from flowering plants. The beeswax in these candles that are lighting up the room is summer light stored by those bees. And you are holding some of that sunshine in your hands—the mead we're about to drink together is made from the honey from the roof.

"And that honey is precious. Our bees flew thousands of air miles to collect the nectar to make it from local flowers. And they don't care where they find them. In the private garden squares of Notting Hill, flowers are growing in some of the most expensive square feet of soil on the planet. But our bees don't need a key to unlock the gates that keep out the riff raff—they just fly over the fences and take the best of what nature has to offer. They're very democratic!

"And the constituency of our bees on the roof extends to the wildflower area of Wormwood Scrubs, where alongside the drones that fly over the walls of the prison to deliver drugs and weapons, our worker bees fly into the prison garden and pollinate the flowers there. And this act of pollination doesn't just help those plants give birth to their young, it encourages the plants to flourish and grow stronger—even in a place of incarceration.

"I believe that most of the men locked up in there are not evil. I believe they have become disconnected. We are lucky enough to have good friends. And every glass of the mead we are about to drink together contains some of the nectar our bees brought home from the blossom in that prison garden. Where they not only helped to deliver beauty to the inmates, but maybe inspired a little hope.

"For the last six months, as the earth has moved us away from the sun, Darkness has held sway with alarming persistence. Austerity persists. The divisions of Brexit persist. Theresa May, Boris Johnson, and Michael Gove all, miraculously, persist. Even Kim Jong Un is still cackling as he chews his Korean kimchi.

"But the local event that overshadows all of them, and literally casts a shadow on the beehives above our heads, is Grenfell. This horrific tragedy, in which so many lost their lives, and so many survivors have been profoundly damaged, this tragedy has revealed the generous spirit of an extraordinary community that was powerfully, quietly vibrant before the fire but is now justifiably inflamed. In the coming months I hope we will all do everything we can to ensure that the causes of Grenfell are exposed to a light so piercingly bright that such a wholly avoidable tragedy can never happen again.

"I know that there are some of us who doubt we can achieve this. Who doubt that merely shining a light on things can cause them to change. Even bees aren't deflected by sunlight—they use it. They use the position of the sun to navigate to the flowers that feed them. A bit like politicians. But, as the bees land on a plant to drink its nectar, they're unaware of the extraordinary, invisible things that are happening as beams of sunlight land on its leaves—the plant is converting solar energy into sugars, chemical building blocks that allow it to grow, and bloom. Light into matter. This is not empty, fruitless debate. This happens. It may seem painfully slow, but light causes material change.

"*Synthesis* is the ancient Greek word for 'coming together,' *photo* is ancient Greek for 'light.' And right now, gathered together in this light given to us by the bees, *we are photosynthesis*. Together, we can grow and blossom. Together is an ancient power. Together we can make change happen. Even on the darkest, longest night.

"So tonight, as the earth begins to tilt us back toward the light, please raise your glasses of sunlight in a toast to the true star of photosynthesis, now returning to lift the darkness . . .

"Ladies and gentlemen, I give you . . . the sun!"

11

Spring Awakening

Bees know with precision when the first of the lengthening days after the winter solstice dawns. Until very recently, like all civilizations whose astronomy was dampened by clouds, our best effort to observe that day when the position of sunrise on the horizon begins to move southward after six months of moving northward was only accurate to the best of three: to our eyes the midwinter sun would appear to rise in the same place for three days before changing direction. The *stice* of *solstice* is from the Latin for "to stand still." And without the benefit of bees' more discerning eyes that can always pinpoint the sun as a source of polarized light, early surveyors marking up the positions of their standing stones would have had their calculations stymied by freezing dawn fogs and winter clouds.

The importance of this day was as massive as the stones our ancestors dragged and stood in their best guess of the positions that would permanently mark it: the dearth of winter was past its peak and the dawning prospect of food was on the horizon. We would celebrate this red-letter day with feasting, waiting out our three days' margin of error until we could be completely sure the New Year had been born: December 25 in our current calendars. Nobody really knows if the bees do anything special to celebrate when winter peaks, sometime between the 21st and 24th of December, but that's when I check that mine have got enough food with a reassuring heft that somehow makes my Christmas pudding taste a little sweeter.

OPPOSITE *Star turn*

The other living things that are also using their photoreceptors to quietly detect the beginning of the lengthening of the days are plants that now know it's time to start growing—the tempo of this phase in their circadian rhythm accelerated by rising temperatures. And as soon as the first spring flowers bloom, your bees will be ready to brighten the landing board of the hive entrance with the reds and ambers of life-restoring pollen. These vivid traffic light colors are your cue to get ready to go for your second annual occasion of activity at the hive.

Harvesting reduced our idle Warré hive by one box. It's now time to put it back, and because that pollen being flown in is going to be turned into babies that need accommodating in new comb that hanging chains of bees will build downward, our new box needs to provide construction space at the bottom of the hive.

The daylight robbery of honey in the autumn might be the zenith of the year for the sweet of teeth, but the twilight subterfuge of "nadiring" your hive in the spring by stealthily adding another box underneath is a high point for the bees. Or at the very least, slightly elevated.

We're teetering on the margins of idleness here: because nadiring unavoidably involves lifting up our idle Warré hive with all the bees inside. The good news is that if we do it just when the pollen and nectar start coming in at the beginning of the year, the hive is at its lightest: the stores of heavy honey we left for the bees to overwinter have been largely consumed, and the population of the hive is close to its lowest.

After the flagrant disruption of the high noon autumn harvest, I try to make nadiring almost imperceptible to the bees. Just before twilight I don my aromatic bee suit and take advantage of the height offered by the chimney breasts either side of the valley roof where my hives are perched: I line up a scaffold pole that rests between the chimney pots, spanning the roof so its center is directly above the hive. There I tie a very basic pulley that costs less than two of those pints of beer in a London pub. I gather my hive tool, an empty, insulated hive box complete with waxed top bars, sufficient nylon cord to tie the hive to the pulley, a water spray, a bee smoker ready to be lit, and I wait until the bees are all in for the night, but there's still just enough light to see.

I gently take the heavy roof off the hive. But before I lift it, with my hands making contact I might casually mention to the bees inside, in an absence-of-harm lullaby kind of way, that I'm about to give them some more room but I hope that this won't disturb them. They've already smelt me—my exertions with the scaffold pole made sure of that.

I flip up the weatherproofing roofing membrane of the lowest box, loop the nylon cord around the handles, rig it up to the pulley, and take up the slack. I gently prize the hive tool between the bottom box and the base to break the propolis seal. With one hand I apply tension to the pulley rope, with the other I get ready to steady the hive: when it's lifted free of the base it can rotate with the twist of the cord, and if this starts to happen I can slow any spin right down until it reaches its static point of happy equilibrium. Then I can use the pulley to *slowly* and *smoothly* lift the hive by more than the height of a hive box. This is not a maneuver the bees will be familiar with— we don't have evidence that trees often levitated like this over the last fifteen million years—so as I perform it I'm not emulating a tree. Instead I'm channeling an event that took place in Paris more than fifty years before timber-related French beekeeper Émile Warré and Parisian obstetrician Adolphe Pinard were even born: on November 21, 1783, the Montgolfier brothers witnessed the first untethered manned flight in the hot-air balloon they had created.

Of course the hive is much too heavy to be lifted by the heat the bees have generated inside it—I am doing the lifting with the pulley— but by making the hive rise as slowly and smoothly as an untethered hot-air balloon, we're preserving as much of the hive's hot air as possible. That lighter air is trapped up inside the hive as it would be in the Montgolfiers' balloon, and if we minimize turbulence at the open bottom of the hive, we minimize disruption to its hot air, along with the pervading scent of the queen, and all the other pheromones it communicates within the hive.

Once maximum elevation has been achieved with minimal awareness, I tie off the pulley rope, remove the mouse guard from the entrance, and place the empty hive box on the base with a gentle star turn to avoid hurting any confused crawling bees that haven't found themselves uplifted. The hive is then slowly and smoothly lowered down toward the empty box. When it's close, I hold the pulley rope in one hand and orientate and align the hive to the box below, using the hovering version of the star turn before lowering it the final eighth of an inch.

Tension and suspense directed by Hitched cord

Readjust the skirts of the weatherproofing and you're done—usually without resorting to either spray or smoker—leaving the snoozing bees to wake up to some extraordinarily convenient extra space.

Of course this rooftop chimney-breast-scaffold-pole-pulley method might be just a little niche. Although there are many types of ingenious lifting aids and systems, it's worth remembering that Émile Warré named his creation the People's Hive and not the Person's Hive. Though you might easily be fit enough to nadir the hive by yourself, in the absence of mechanical advantage, it's good to have a friend to share the burden.

Either way, because your hive may not be suspended from a pulley, you will need somewhere to put it whilst you're adding the empty box to the base. That might be in the arms of a strong friend, but any delays you encounter might not only strain both your friend and your friendship but also increase the possibility of bees falling out of the hive wherever your friend is standing, and flirt with an unintentionally flat-footed encounter with the queen. So I'd recommend using another empty hive box as a temporary stand—rest it next to the hive on the Correx tray you made to catch drips of honey during harvest, but now can catch falling bees. You can even lay an old sheet underneath, which will show up any bees that land outside the tray's catchment area, even at twilight.

Hovering, bee-nudging star turns are difficult to accomplish when you're lifting the hive by hand—you're so close to it, and it's so much taller than the single box you harvested that you can't really see the edges of the boxes between which you and your slightly too close friend are struggling to avoid crushing the bees. One way to get some distance and improve your perspective is to convert your hive into a simple, temporary sedan: two sturdy lengths of two-by-two-inch timber temporarily held in place under the wooden handles of the lowest box with some silent, stealthy screws won't make the hive less heavy, but they will allow both of you to see what you're doing when you lift it like porters carrying a sedan chair.

Even with this improved vantage, star turns are best avoided: if you try shuffling to rotate the hovering hive, you will discover that it takes more than two to tango—it additionally requires a whole lot of very precise coordination that you and your dance partner may not have rehearsed to perfection. Safest is to stand your ground, align the boxes, and lower the hive very, *very* slowly through the last few moments before contact. Imagine a bee feeling a box bearing down on its head, and agree with your friend to give it the time to either crawl in or out. And then agree to slow down even more—remembering how meanderingly dozy both of you can be when you're half asleep.

If any bees have landed in your temporary stand, move it close to the hive entrance and they will return under their own steam to its familiar warmth and pheromones for the night. Perfect is the enemy of good, especially if things have taken longer than you'd planned and it's now so dark you actually can't see what you're doing. So as long as the newly extended hive is intact, walk away and leave any unfinished tidying up till daylight returns—the bees won't care if their bigger hive is still attached to a pulley or looks like a sedan chair for a wee while.

I used to sometimes repeat this process later in the season to accommodate particularly vigorous colonies, nadiring additional hive boxes in response to their prodigious, precipitous construction of comb. And there's something awe-inspiring knowing that the six hive boxes you're lifting up to add a seventh are all brimming with energetic, healthy, purposeful bees living in a hive that now stands taller than you. But it's not just that those six boxes are fantastically heavy and the seven totteringly unstable that's made me retrench to a four-box maximum: if a colony is so vigorous and its food supply so plentiful that it can generate enough bees to populate six boxes, I'd rather encourage it to become two colonies that each fill three boxes in their own separate hives—with two queens rather than one—spreading the risk from queen death and increasing genetic diversity.

But of course the neat, easy mathematics of such asset division and reassignment are the grist of our superficial spreadsheets. For the bees, it's murder—literally.

Shortage of room drives upsizing. We used to love our one-bedroom apartment—so much so that it took the arrival of our second child to precipitate moving to somewhere bigger. And though there are many factors that influence a colony's decision to divide in two with half the bees swarming off, lack of space is pretty non-negotiable—the bees will already be inhabiting every available nook and cranny with the most ingenious efficiency. So very quickly our vigorous bees will need to look for more space for their rapidly increasing number of babies.

And we can provide it in the form of the most desirable bait hive that we make available pre-rhyme: although May is traditionally the time for early swarms, if a colony is set to burst at the seams, scout bees could be house hunting from early April. We want to do everything we possibly can to tempt that vigorous swarm to move in with their queen. Now is the time to go beyond "bait" and deploy the oldest catchphrase in the real estate agent's lexicon: "original features" are persuasively popular among prospective purchasers, and bees are no exception. For the top box of this bait hive use the propolis and wax-lined hive box you harvested from this colony the previous autumn complete with its previously loved top bars. It's the ultimate, personalized original feature. To the scout bees it won't just smell like home, it will smell *identical* to *their* home . . . but empty! And the perfect size!

We know that the size of the two-box bait hive does indeed seem to be just right because in his experiments to determine this, Professor Tom Seeley offered scouts a choice of bigger but otherwise identical bait hives, which they declined. Smaller but identical were declined too, so two boxes appears to be Goldilocks.

And it's likely that just as bees weren't anticipating that their tree cavities would spontaneously levitate at twilight, scout bees wouldn't factor into their calculations that an ideal cavity would miraculously expand at some later moment in time—just when they needed extra space. Like the door to that broom cupboard in our dreams that magically opens onto a crystal-chandeliered ballroom. It appears the scouts don't have ballroom envy—the broom cupboard's all they need.

So if the volume of two boxes is the space they naturally choose, why are we giving them the oversized dance floor of four?

We're giving them a bigger home than they would naturally choose so we can take their honey without killing them. If our working hives were only two boxes big, harvesting half the hive—one whole box—would be draconian, and harvesting individual combs requires disruptively cutting them out with the Warré hive knife— if you're lucky enough to find an individual comb that doesn't have babies living in its lower half.

Even though we're trying to be as considerate to the needs of the bees as we can, the four-box hives that allow us honey also begin to take us down a path that leads to farming: like extending the lactating period of cows beyond the direct suckling of their calves, and prolonging the egg-laying of hens beyond spring—the specific time wild birds procreate.

GLOBAL WARMING CAUSES DRAMATIC REDUCTION IN SPARROW AND BLUE TIT EGG HARVEST. We don't typically read headlines like this because the thousands of people in the United Kingdom who keep bird boxes do so without any expectation of consuming their eggs.

So if you're uncomfortable with the consequences of your honey consumption, why not keep a honey bee box the same way? With the ten-gallon volume of two Warré hive boxes, it would be much bigger than a bird box, and when fully occupied much heavier, but apart from that it would operate on the same principles: like birds, the honey bees aren't forced to move in, and if they choose to inhabit the box, they are also free to leave. It offers them protection without obligation, and without human management—you leave them alone. There is no honey for the bee-box-keeper, it all belongs to the bees. They'll know what to do with it: they've had millions of years of practice with surpluses. The benefits to the bee-box-keeper are also similar to those for the bird-box-keeper: helping a beautiful wild animal connects you to your local environment more deeply than you might imagine—you now have a bee in the fight.

At the other end of the technological spectrum of minimum disturbance of the bees is the Flow Hive. It allows the beekeeper to drain honey out of the hive without the bees even knowing it's happening: mechanical levers and cams open up plastic honeycombs at each of what the bees assumed were hundreds of solid, honey-tight hexagonal cells. The honey they contain drips out and then the cells are closed back together to be refilled by the bees. "Honey on Tap" is a selling point, but it distances the beekeeper from the bees: its design allows less engagement but more exploitation—empathy outsourced. The system is a technologically advanced, much more expensive version of the Langstroth frame-based hive and has to use a queen excluder so its mechanism is dealing with exclusively honey-filled cells. And even though the bees will coat these plastic cells with the minimum of propolis, only time will tell what that sticky brown substance will do to a mechanism so intricate you'd need to be a watchmaker to make one, not just a cabinet maker. And though wax comb resonates at a bespoke bee-friendly frequency, the plastic comb doesn't and may distort the bees' vibrating communication: a plastic violin doesn't sound much like a real violin, let alone a Stradivarius. Although it might appear to be the idle beekeeper's dream come true, for me it strays uncomfortably into the territory of conventional bee-farming turbocharged with idle honey extraction. If quality, local, raw honey is your main priority, make friends with a local beekeeper: it's much cheaper. If "Honey on Tap" is your goal, make friends with many beekeepers.

Bees in my four-box hives will tend to run out of space and need to swarm more frequently than bees in a six-box hive. But because bee-box cavities the size bees naturally choose are even smaller and would promote even more frequent swarming on that basis, maybe we should consider this increased frequency normal.

Conventional beekeepers actively suppress swarming so they don't "lose their bees," and there is a risk that your swarms may not choose your no-brainer bait hives, but my experience has been that they are so keen they don't even bother to form a cluster—every single swarm that's moved into my bait hives has always flown straight in from its original hive without further ado. Sometimes they don't even fly and appear to walk down the front wall of my house, tenaciously treacling down from the roof all the way to their chosen bait hive on the tiny balcony outside my office window: a nuisance-free example of good "swarm hygiene" from my neighbors' and any passing pedestrians' point of view.

And there are benefits to frequent swarming beyond the spreading of genes and risk. For a brief period after the swarm has departed the old hive and moved into the bait hive, there are no babies in either location. This break in the brood cycle is a kick in the teeth for varroa mites—the essential reproductive phase of their life cycle involves the female mites sneaking into a cell just before it's capped to allow the bee larva inside to safely pupate. To avoid detection the mite hides in the pile of brood food the bees have stashed in the cell to see their youngster through to winged adulthood. Varroa mom lies in wait as unwitting worker bees seal her in with *her* babies' food supply—namely, the baby bee. No baby bees, no baby mites.

Even when a swarm moves straight out of its hive and into the bait hive, and even if the bees managed to construct a cell and the queen laid an egg in it that same day, there are still at least nine days when that larva is being fed in an open cell before it's capped: for this period any varroa mites that have hitched a ride on the bodies of bees in the swarm have no opportunity to reproduce.

Back at the original hive, it was a similar nine days to the capping of a different cell with a specially selected larva that triggered the release of the swarm. Whilst the old queen was being slimmed down to make her light enough to be able to fly, worker bees constructed a special wax cell.

It's bigger than all the other cells to accommodate the egg of a new queen who will grow to be larger than all the other bees. The workers place an egg in the queen cell, and when it hatches into a

larva they feed it a slightly different diet to other bees. Along with the bee bread fermented from pollen, all bees are fed a little royal jelly—a milky goo secreted from glands in the mouthparts of nurse bees. But the specially selected larva in the queen cell gets fed lots of it and it makes her develop differently—into a queen.

On the ninth day the workers seal this queen cell with a cap of wax, and leave her inside for the seven days it takes to pupate from larval princess into virgin queen bee. As soon as this cell is sealed, and succession assured, the old queen leaves the hive with her swarm.

But it's a succession selected by the colony, not the old queen. It's the workers who choose the egg that they give the special royal jelly prerogative. To our eyes the chosen egg looks indistinguishable from all the others. But the bees choose from around 1,500 female eggs freshly laid that day by the old queen. And their genetic survival depends on a wise choice—the new queen will be the singular proponent of their DNA—so they select very carefully.

Worker larvae and eggs

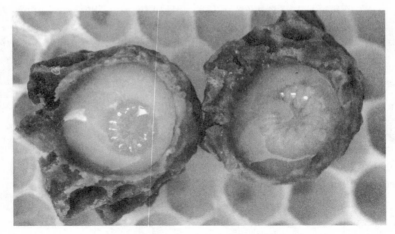

Princesses awash with royal jelly in palatial queen cells

And to make it more interesting, the workers hedge their bets: they build and fill a new queen cell every one of those nine days between the first selection and the old queen leaving—sometimes more. So by the time the old egg-laying queen is gone, there can be about ten potential new virgin queens growing in their cells in the hive.

Commercial queen breeders find it easy to trick worker bees into raising eggs as queens by placing eggs in artificial queen cell cups that they insert into a hive. These cups can be made of plastic so long as they are the same size as the real thing—the workers feel the grandeur of the accommodation and deferentially oblige by providing the required amounts of royal jelly to the larvae the breeder has placed there.[10] Move over Thomas Cromwell, this isn't just about providing a few wives for King Henry VIII, this is industrial queen making! And it services an industry with an increasing demand for royalty.

When a queen dies, a quick way to resume business as usual in the hive is to replace her with one you can have delivered in an envelope the next day, instead of having to wait at least seventeen days for the bees to fabricate a wax queen cell and raise a new queen from an egg. If you simply put the queen you've paid for into your hive, she'll be attacked and killed by your bees because she smells foreign and wrong. So she comes in a little queen cage with a few bees from her retinue to look after her. Protective mesh stops your bees from getting at her, but allows her scent to get to them. The exit from the cage is blocked by a plug of sugar fondant: by the time this has been eaten through by the bees, the new queen smells familiar, wonderful even, and they're ready to welcome her onto the job.

This mechanism for queen replacement is so convenient and quick, industrial beekeepers can use it to retire a flagging, unproductive queen, or even a queen that's producing inconveniently defensive bees that don't like the frequent intrusions of the conventional beekeeper and exhibit aggressive traits.

---

10   Anecdotally, sometimes the bees reject the egg that's been placed in the artificial queen cup and replace it with one they choose themselves.

But queens are living shorter working lives than they used to and maybe this is the price of our intervention for profit. Queen breeders don't throw away healthy queens, they sell them. But in the wild being healthy isn't enough—at any price.

Inside your hive there is a further process of selection that would be brutal on breeders' profits and drain the blood from the cheeks of even Cromwell and King Henry: only one of the ten new virgin queens can be Her Majesty, and on hatching she will immediately try to slaughter all her royal sisters by stinging them to death.

We all know that when bees sting us they die, but our skin is an anomaly: their stinging mechanism primarily evolved to attack other insects, again and again. And they will spare no venom in this struggle for supremacy. Drawn by the reek of competing queen pheromones seeping through the hive, the first virgin queen to hatch rushes around the hive looking for queen cells, puncturing the protective wax barrier with her sting, trying to kill her sisters while they're still inside and can't fight back. But she may not be the strongest or the cleverest. And while she is engaged in mortal combat with a sister already hatched, others bide their time in the relative safety of their cells, poking out their long tongues through a tiny hole they have made in the wall to receive energizing honey from their retinue, while they listen to the death throes of their embattled sisters resonating through the wax of the honey comb. And calculate the best moment to make their bid for power . . .

They are aided and abetted in these calculations by their retinues of worker bees who, in response to the shifting political landscape, exert a high degree of control over the emergence process, including vibrating her cell, obstructing the virgin queen's exit, and resealing holes she's eaten through the wax cap to tell her "Not yet!" Some of the virgin queens who have a large enough retinue might choose discretion over valor and leave the hive with a small "cast" swarm that might be lucky enough to survive. But all the sisters that remain will be killed save one: the new virgin queen.

Close to nine out of ten of the queens that commercial breeders sell wouldn't have survived in nature: they would have been found inferior to a sister and died for it. Commercial breeders introduce these inferior queens into colonies and breed from them without selection. And they do so with their daughters and their daughters' daughters. This has been going on for decades, and the world is now full of bees from queens that would never have survived to give birth without our intervention. But even if it weren't financial suicide to sell only the best queen out of ten, how would the breeder select her? By what criteria would they be able to judge?

You are spared these harsh eugenics by a very special female: the one-and-only surviving virgin queen in your hive was selected by experts from fifteen thousand candidates freshly laid by her mother, and then ruthlessly self-selected from ten expertly chosen sisters. Now that the sororicidal bloodbath is out of the way, the victress gets to deal with her virginity.

She's escorted by a posse of her retinue to the imaginatively titled Drone Congregation Area: a place in the sky chosen for reasons best known to the drones who congregate there. It's where the boys hang out, hoping, as ever, for sex. The virgin queen wafts into the midst of them, trailing pheromones that irresistibly announce her immediate availability, and flies straight upward, as fast and as high as she can. Desperate males give chase, attempting to catch her before their competitors do. As the strongest and most skillful drone gets close, his vast endophallus looms out from the rear of his abdomen, looking more like the lunar landing module than the dong of Dirk Diggler. The virgin queen will avail herself of him and then rip off his genitals with a snap that can be heard from the ground, to which he falls. Life goal and death in the same instant, if he's lucky.

Typically the queen will mate with, and kill, between ten and twenty of the best drones available before returning to the hive—with the last set of male genitals hanging out of the back of her abdomen in shared celebration of a successful outing, like bunting by Hieronymus Bosch.

Replete with enough semen to last her a lifetime, she's had it with sex, and her time for choosing is over. When she starts her daily royal duty of producing hundreds of eggs, even though at the moment of laying she uses her bladder of stored sperm to selectively fertilize the eggs into females, or to lay them unfertilized as males, it's the colony that decides the relative numbers, not her.

But she won't start laying until after the last of her mother's eggs have hatched. And as their numbers still pupating in their capped cells diminish to zero, so does the opportunity for varroa mites to have children. Even if the new queen starts laying the second Mom's last baby hatches, the number of nourishing cells available for baby mites has crashed from as many as two thousand to zero in fourteen days, and will remain at zero for a further nine days until the new queen's first child is sealed in to pupate.

It seems that mother and daughter queens are quite comfortable with their hives being close to each other. And the sight of distinctly different colors of pollen going into each hive, combined with visibly diverse behavior at each entrance, confirms that their independence survives proximity.

I was reclining on my roof one October, mesmerized by the beauty and subtlety of these revelations of the different characters of mother and daughter colonies, when something appalling began to happen: bees were being evicted from the hive against their will. No matter how desperately they tried to cling on, dozens were ganged up on by other bees and forcibly thrown out to a certain death. This is not the targeted political assassination of rival virgin queens that occurred under cover of darkness inside the hive to determine royal succession in the spring. This is genocide, or more accurately, gendercide—all the males, even the good-looking ones, are being summarily excluded from the colony, and they are being killed by their sisters. Casanova and Falstaff discovered that however charming, sexy, or fun you are, there comes a time when your indulgers question your usefulness. Cost benefit. And at this time of year the bees are focused on preparing to survive the cold winter. Not flirting. No room at the inn or the table for anybody not pulling their weight . . .

When I first saw this "Drone Slaughter" at a hive entrance I immediately took it personally. As a honey-stealer more than as a male: if I'd left more honey in the hive, perhaps the drones could have been indulged a little longer? Survived the winter? Had I caused a carbohydrate shortage that drove sister to kill brother?

But then I saw that the gendercide didn't happen at every hive. And pondered the huge and uncharacteristic waste of hard-earned, lovingly reared protein and fat that chucking out all that healthy beefcake represented. Some male bees are useful to have around long after the early summer swarming season when most virgin queens mate. If an established queen dies or loses her vigor, she will be superseded—workers will raise a virgin queen, and this virgin may need to go on a mating flight at any time: she needs an immediate supply of ready and willing drones so she can quickly mate and return to the hive to carry on the egg laying whilst her mother is unceremoniously dispatched for the good of the colony.

And those colonies that regulate the supply and demand of drones with beautiful accuracy have no surplus to evict at the end of the year. That's the goal of all colonies, but sometimes the demands of a shifting environment foil the best laid plans, and in terms of disseminating its DNA, it's better for the colony to err on the side of producing too many drones rather than too few.

Not much comfort for those individual males now dying without ever being "lucky" enough to catch a queen. They could be forgiven for questioning the purpose of competing with other males: like buses, the colony ensures that there will be another along in a minute, so why bother? Can any male really make his mark?

When the female workers carefully select those eggs they'll raise as virgin queens, all of them contain the queen's DNA so there's nothing to differentiate them on that criterion. But they were fertilized with sperm from the queen's mating flight. So they are likely to have different fathers—from the random mixture in the sperm bladder. But if by chance these eggs all had the same father this might be a disaster—what if he'd been the worst of the bunch of those successful drones? The virgin queens might fight it out and select the best of themselves, but from the lowest paternal base.

There would be an advantage in making sure that every possible father had a virgin queen in the fight—so some potentially brilliant individual DNA wasn't missed. Could a species successfully survive so long without optimizing this? Maybe when the workers select those special eggs to be raised as queens they don't just pick robust, healthy specimens, they also make sure the selection of eggs represents the best of those lethal sexual encounters. We know that worker bees can identify genetically defective eggs by smell and kill them before they develop—could they also smell who the egg's dad is and select accordingly?

I put this hypothesis to David Tarpy, professor of entomology at North Carolina State University, who's working on things like how honey bee colonies regulate queen reproductive traits by controlling which queens survive to adulthood. David replied,

> Larval recognition cues and even pheromones have been discovered, with the diploid drone "smell" being one of the most obvious. To date we don't know how (or even if) workers are detecting eggs that are sired by different drone fathers. As far as how selection might be acting on this, you describe my major line of thinking about this.

So science may yet answer this question. But even if some proud dads have a chance for their daughter to become a queen, we must remember that because male bees come from unfertilized eggs, and have no fathers, the only DNA they can bring to the table is their mother's. And their mother was the issue of a union between a queen and a successful drone who also had no father, and brought no male DNA to her. He too could only contribute his mother's genes, which are those of a successful sister-killing queen, descended without any male genes from a dynasty of sororicidal queens that have been succeeding for millions of years.

But not entirely without the help of males. At the end of the mating flight he was born for, the moment his lunar landing module reaches its destination is the culmination of a journey to carry life to the farthest point of its reach.

To boldly go. Refueling at any and every hive along the way, to deliver the diaspora of his queen's DNA.

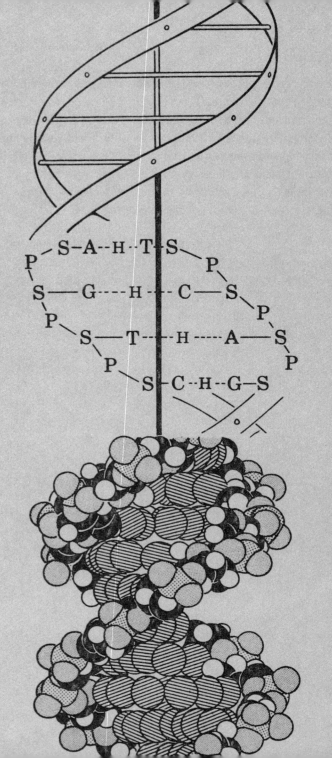

12

Seize It!

Two blokes walk into a pub and say they've discovered the secret of life.

Only on February 28, 1953, they weren't joking. The alcohol, whose intoxicating whiff permeated James Watson and Francis Crick's announcement that they'd discovered the double helix structure of DNA, had also been an essential solution to the problem of extracting it. Three years before those celebrations at The Eagle pub in Cambridge, Swiss scientist Rudolf Signer came to London with a half ounce of some white fluffy stuff from his laboratory in Berne. Signer had developed an alcoholic technique to extract the DNA from the thymus gland of calves. Inside the cells of calves, bees, humans, and all living things, our genes come packaged as chromosomes: DNA makes up about half, and the rest is a protein coating that surrounds and protects it—and obscures it. Signer used alcohol to rinse off this protein coating and he did it so gently that the underlying structure he revealed was pristine: the DNA of Signer's calf's thymus was the dog's bollocks.

After his talk to the Faraday Society on May 1, 1950, Signer shared those few grams of calves' DNA with his fellow scientists. One of them was Maurice Wilkins, who took it back to his boss Rosalind Franklin, and in her lab at Kings College London they used techniques she'd developed in x-ray crystallography to photograph it. One of their beautiful portraits—*Photograph 51*—was a lightbulb moment for Watson and Crick who, along with Wilkins, were awarded the Nobel Prize in 1962. Neither Franklin nor Signer received the same

OPPOSITE *Basic behavior?*

kind of recognition for their work on DNA, but beyond gender and prejudice this shows how fallible we are at recognizing what's really important: in 2012, Sir John Gurdon won the same Nobel Prize for his work in epigenetics. The prefix *epi* means "on the outside," and epigenetics is the science that goes beyond the structure of DNA and studies how genes actually *work*. Now the award-winning discoveries are all about the behavior of that same "epi" protein coating we'd been throwing away for forty years—casually rinsing it off chromosomes so we could get a better look inside.

That same year that epigenetics was recognized by the Nobel committee, researchers in the United States played a cunning trick on honey bees. While foraging bees were out of the hive gathering food from flowers, the scientists removed all the nurse bees that were looking after the babies inside the hive. When the foragers returned with food to find that nobody was looking after the kids, half of them quickly reverted to being nurses and took care of their little darlings. The other half carried on gathering the essential food. All the foragers had previously been nurses as part of their career development, so they knew how to do the job, but it turned out the change in role wasn't just a change in the bees' behavior—they'd also changed their genes.

The nurses and the foragers share the same inner DNA double helix of Watson and Crick fame. But their DNA protein coatings are different.

DNA is a set of building instructions—how to build a human, how to build a bee. Like an architect, the DNA provides the plans—it doesn't do the building, or maintain the building; other specialists are contracted to execute the construction—such as plumbers and electricians. And it's important that the plans don't get lost or damaged. So if the building were a tower block, imagine its DNA double helix as a complete set of architectural drawings that describe every aspect of the edifice, safely organized in a planning chest with dozens of drawers, each with a lock. These drawings don't do anything. However long you watch them.

Next to the planning chest is a photocopier, so our plumbers and electricians can unlock the drawers of the plans they need, make a working copy to take on-site, and return the original plans to the drawer for safe keeping. But although many of the plans needed by plumbers and electricians are the same—room function and layout—some are different. So only plumbers have the key to the drawers with the piping diagrams, and only electricians have the key for the drawers with the wiring details. To help the right plans get to the right tradesman and prevent your light fittings from drenching you with water and your shower sizzling with sparks.

This locked plan chest is the equivalent of the epigenetic protein protectively coating the precious DNA. This protein coating is divided up into "drawers" with tiny chemical "locks." Scientists call these drawers Differentiated Methylated Regions and these regions encompass different sections of DNA—different sheets of the "plans." But by themselves neither the plan chest nor the keys move. They don't do anything. Nothing happens until a plumber or electrician turns up. But the plumbers and electricians I know aren't ardent photocopiers: I've never heard one enthusing about a wiring diagram so beautiful he wanted to make a copy to just take home and hang on his wall, or piping plans she wanted to print on a T-shirt because she believed in them so much. That's not why they come.

Our locked plan chests get visited and opened up when a frozen pipe bursts, when a power surge fries a fuse box, or when all the nurse bees have disappeared: a change in the environment is detected and messages sent to precipitate action. Now plumbers or electricians or foragers arrive and act on the instructions they copy from the plans to restore normal function.

The scientists who spirited away all the nurse bees measured fifty-seven of these epigenetic Differentiated Methylated Regions that were specific to nurses—their inner nurse "key" had unlocked those fifty-seven drawers containing the specific nurse instructions all bee DNA contains. When the foraging bees flew home, they had no access to those nurse instructions—they were locked because why would foragers need them? They're pilots now! Like Amelia Earhart! With inner pilot "keys." But when they discovered the babies untended, half of those pilots unlocked those fifty-seven regions of nurse. They weren't just behaving like nurses, their reaction to the situation gave them the "key" to unlock their inner nurse. They *became* nurses, genetically.

Virgin queens aren't born, they're created by diet. Inundating a female baby with extra royal jelly doesn't just fatten her up; an environment of so much milky goo unlocks the royal regions of her DNA. All female babies have these royal regions, but only a few are chosen to be exposed to an environment that unlocks their expression and makes an ordinary baby become a physical queen. With a fight on her hands.

We often think of genetic change taking place very slowly over thousands of years, but epigenetic change can happen very quickly: those foragers that turned into nurses needed to be feeding their untended babies within hours. And they needed to know how. But it would have been disastrous if all the foragers had *all* responded by becoming nurses—the colony would have starved without the food they were bringing in. So a forager responding when 49 percent of her fellow pilots had become nurses would know how she must join them, but a forager responding to 50 percent would know she needed to remain a forager. Epigenetic changes made by individual bees happened in rapid response to a quickly evolving group dynamic.

And if they had been able to achieve all this merely from memory, why the measurable epigenetic change? When we say that "you never forget how to ride a bicycle," where is that memory stored? If it were stored in your DNA—you were born with the latent ability to ride a bike without consciously knowing it—the sight of a bicycle might unlock your cycling genes in time for you not to fall off. But we have large brains that currently believe they possess swiftly retrievable Random Access Memory like computers, so we think we store learned cycling information there rather than in our genes. But maybe bees, with brains the size of a grain of salt, utilize the information storage capacity of their DNA in combination with nifty epi-genetics to be able to recall behavior quickly and accurately.

This might be particularly useful for behavior that is rarely, if ever, required: storing the information for how to react to thousands of different environmental circumstances that are very unlikely to occur would place an unnecessary burden on the tiny bee brain and be better stored somewhere safely unforgettable, like the chemical structure of the DNA that is present in their every cell. In their life-time, the chances of a forager returning home to find all the nurses gone are similar to returning home to the smell of smoke: close to zero. Most bees will never live to experience a mass disappearance of nurses or a forest fire and they've not had drills in either, but they know what to do.

To our eyes the bees' response to smoke looks almost instanta-neous, a reflex action in response to the smell: they recoil. But the col-lective retreat into the middle of the hive designed to give them the most insulation from the heat of the fire, and filling their huge crops with honey as fast as they can to increase their body mass to make them take longer to be heated up by that fire, are both responses to buy them time—increase their chances of still being alive when the inferno has burned itself out. Bees haven't been taught these responses, and their sophistication begs epigenetic instruction.

Every bee knows how to safely build high-rise structures of wax comb. But that is the only thing bees build. They don't build fire engines or bridges or smartphones. They don't know how. We do, but the diversity and complexity of the millions of things we build is so vast that we cannot hold all the necessary construction plans in our heads, so we download them somewhere safe where we can retrieve them at will.

There is an apocryphal anecdote about Albert Einstein: traveling on a train, he saw the ticket collector approaching and started rum-maging in his pockets. The ticket collector recognized the famous genius, and seeing the look of consternation on Einstein's face, told him that there was no need for him to see the ticket, he trusted him to have paid his fare. Einstein thanked him, and the ticket collector carried on. But when he turned and looked again he saw the great physicist down on his hands and knees still searching for his ticket. The ticket collector went back to reassure him: he knew who Einstein was, everyone knew, he was sure he'd bought a ticket. Einstein replied, "Young man, I too know who I am. I just can't remember where I'm going."

Einstein had downloaded responsibility for the memory of his destination to the ticket on which it was printed. The disadvantage of this strategy is that you have to refer to your ticket every time the train arrives at a station in case you miss your stop, but the advantage is that even in a country where you don't speak the language or even understand the alphabet, you can silently share this memory with anyone who can read it, or by yourself you can visually match the shapes of the letters to the image of those displayed on the platform of your destination. And at that moment, the data retrieved becomes information that makes you act: you get off the train.

In 2017 an aspect of UK government went digital and ended a tradition that stretched back more than five hundred years. Acts of Parliament, the democratically agreed laws of the land, had been stored in writing so their data would be clearly defined and citizens wouldn't forget them. And to ensure that these outsourced memories would last as long as possible, they were written on a protein coating: not the molecular epigenetic one Rudolf Signer rinsed from the DNA of calves' thymus glands—we'd upscaled to the whole calf and wrote our laws on its skin. Vellum parchments inscribed with eight-hundred-year-old signed copies of the Magna Carta are still in good shape. When they're 1,600 years old we'll know if the digital alternatives in which we're now storing our laws are half as good.

In 715 CE, long before lending libraries or printing, and five hundred years before the ink of the signatures had dried on the Magna Carta, a religious leader in an abbey on the tiny tidal island of Lindisfarne off the coast of Northumbria in Northern England was making a copy of his god's Holy Scripture so that its word could be spread:

and he did it with the help of the bees. Local people would have had very little idea of what he was up to inside his scriptorium, but they would have noticed an increase in beekeeping to meet the voracious demands of Eadfrith, the Bishop of Lindisfarne, and the best of it would have been that beeswax was his proclivity—he wasn't interested in honey or mead. Though there are no records of a boom in either over this period, perhaps everyone knew to keep very quiet about such a sweet bonanza and not disturb the bishop's work: Lindisfarne mead is legendary and still made today.

The religious scribes of the Dead Sea Scrolls had used the same calfskin parchment five hundred years earlier, but because he wanted to reach the vast numbers of souls as illiterate as the bees, Eadfrith illustrated his manuscript with lavish designs so his flock could see an enhanced interpretation of the teachings of the gospels with their own eyes. And the priests that read the word of their God out loud to them could also be inspired by its beauty, vibrant on the page before them. His illustrations were intended as a visible blessing, and to look at them was to be blessed.

Calf vellum was extremely precious and costly. And like a legal document, any evidence of an error covered up carried the implication of tampering—Gods don't make mistakes so believers wouldn't expect to see any in the physical manifestation of their Word. So Eadfrith had to get everything right on all the 150 calfskins the book required. First time. Such a stricture might have dampened the creative ambition of a lesser monk, but for the Lindisfarne Gospels not only did he use 105 different pigments where most illustrators used three, Eadfrith developed a revolutionary method of illustrating and this was where the bees came in.

Traditionally the outlines of the designs would have been scratched out on reusable wax tablets or disposable slates before committing final versions to the vellum by scratching the precious surface of the calfskin. Apart from leaving crude traces of the hand of man in the Word of God, on a pragmatic level when the illustrator was applying pigment—coloring in—he was painting over the only true, final copy of the original design—he was obliterating the downloaded memory and had to rely on his own.

Eadfrith invented two things that enabled him to solve both these problems. The first was a kind of prototype lead pencil. He used this to draw his designs on a small piece of working vellum he would reuse, allowing him to rub out any mistakes. When he'd finalized his design, he'd place this working vellum on the back of the relevant blank page of the book and rub the two together, so a copy of his penciled design was transferred to the book in reverse image. He then fixed the page on top of his second invention: a light box. Now he could see his design projected through the calfskin onto the front side of the page, the right way around. The light enabled him to paint his design without scratching any marks on the front of the page, and he still had the original preserved on his separate piece of working vellum to refer to as paint covered up the image on the page.

This illuminated illuminated manuscript was made possible by the bees: Eadfrith's light box was powered by beeswax candles. He needed a reliable, consistent, and plentiful source of light and beeswax: the book took him ten years to make and is 515 pages long[11]—a page a week is an extraordinary feat, and he could not have accomplished this with variable daylight coming into his scriptorium through windows not designed or positioned for his purpose, especially in the long hours of winter darkness. He had to borrow the sunlight the bees had shaped into the most exquisitely designed accommodation for their young so he could fashion his meticulous designs to illuminate the souls of his flock. And he included every style of illustration to appeal to the widest congregation, like your bees energetically fanning scent trails from the entrance of the hive in an attempt to guide to salvation all who are lost.

But whatever Eadfrith painted, as each of the pleasantly scented but eye-wateringly expensive candles burned down, its light and

11   You can see some of them here: http://www.bl.uk/turning-the-pages/ ?id=fdbcc772-3e21-468d-8ca1-9c192f0f939c&type=book

heat dispersing beyond his endeavor and out into the universe, the laws of entropy held sway. Outside his scriptorium the sun, whose stored light he was borrowing from the bees, was burning like those candles, and the time will come when it too will burn out and become another celestial cinder in a cold, dark void of space. Along with all the other stars.

The second law of thermodynamics that got us wrapping our idle Warré hives in yet another protein coating, wool, is ruthlessly insistent. It will get its way. Entropy says so. We observe our universe and we can see that these laws govern it, but there is a phenomenon they really struggle to explain.

Life.

In an environment where chaos and disorder prevail with mathematical predictability, a self-organizing, energy-gathering entity that repairs and reproduces itself is an anomaly. A maverick resistance to the relentless flow of entropy, life can even become so self-aware as to consider itself intelligent, but still be bamboozled by what, if anything, it all means.

Biologically life can be described as occurring when an organism has a higher rate of anabolism than catabolism—it's building and repairing faster than it's wearing out—but this feels superficial. I've only ever been with one person at the moment of their death, but although their physical body seemed pretty much identical ten seconds before and ten seconds after the person died, the permanent departure of their consciousness in that time was immeasurably the biggest change I've ever witnessed.

It would be embarrassing to discover that in our search for the essence of all life, not just the human version, we consider consciousness the way we used to consider the protein coating we rinsed off mitochondria to get to the DNA we consciously wanted to measure. Before we hit on epigenetics and stopped pouring it down the drain.

If life energetically defies entropy, is consciousness epi-chaotic? Mine certainly feels like it, struggling to make sense of the world, living in a constant negotiation with chaos and disorder. When I started beekeeping I'm sure that a big part of its attraction was spending time scrutinizing a living microcosm of order in the hope that some if it might rub off in the deciphering. The activity of the super-organism of the honey bee colony in its hive superficially seemed so frenetically chaotic: and yet those perfect hexagonal honeycombs; the navigational certainty of bees flying back into the hive laden with pollen from who knows where; the street-filling random blizzard of bees all moving into a tiny bait hive together in a matter of minutes. Could understanding the bees' secrets unlock clarity of purpose and productivity in my own life? The more I've learned, the more I've realized that any progress has not come from unlocking but from throwing away the keys.

Our attempts to corral bees into the service of our intensive agriculture for the pollination and honey they provide for us are tee-tering on the brink of sustainability. Every year in the United States the bee brokers only find enough healthy bees to send on the grand pollination trucking tours so vital to our food supply by the skin of their teeth.

The same global pressures on bee populations caused by loss of habitat, pollution, and pesticide poisoning are also felt in China where the Food and Agriculture Organization found that between 2000 and 2014 honey production increased by 88 percent in spite of the number of hives only increasing by 21 percent over the same period. The additional honey was made possible by humans doing some of the bees' work for them: not out of kindness but to maxi-mize profit by perpetrating fraud.

Honey was harvested when it was watery, long before the bees had safely capped its combs, and on an industrial scale highly fil-tered to remove all traces of pollen. This made it impossible to trace the honey's origins when it was exported—like all intensive farmers Chinese beekeepers were using drugs to keep their stock alive, but in China they used antibiotics like chloramphenicol, which is banned in Europe and the United States because it's carcinogenic. The fil-tered, watery "honey" was then industrially dried, fraudulently sold through third-party countries, and adulterated what consumers all over the world believed was a natural product.

If agricultural subsidies artificially keep the cost of sugar so low relative to the margin of profit that can be made by turning it into "honey," it is not fanciful to imagine farmers contriving to reduce the bees' involvement in honey production to its bare minimum: a sugar solution quickly passing through a bee's crop before she deposits it in some kind of honeycomb for the minimum time necessary to give it a honey-like flavor. Honey farmers would do all the rest of the work on an industrial scale. This might allow bee sheds like those in which we confine battery hens—where the bees are fed cheap, subsidized soy-based pollen substitutes and sugar syrup, their work cycles controlled by artificial light, pharmaceuticals liberally administered to keep them alive, and no need to encounter the sun or a flower. As with battery hens, any flying would be a waste of energy and minimized.

Is this the kind of honey we want? The kind of unsustainable world we want? Because these are choices we can make: choices that add up to cumulatively define who we are.

I started out intending to become a good beekeeper. Now I like to think of myself as a hive keeper, and the distinction is not merely semantic. I don't keep bees—they keep themselves, far better than I will ever be able to imagine. They are wild animals, free to come and go on their own terms. But like an innkeeper trying to entice customers to come to his inn, I can try to provide accommodation that suits my guests. If I have no idea what it is they need and want, I cannot succeed. But the more I begin to understand the bees' changing needs, deliver the right support at the right time, but otherwise allow them to develop according to their own natures, the more likely they are to thrive and stay.

The innkeeper's approach to the bees, handing over keys to your guests rather than wishing you could lock them up, has in some microscopic way connected me to cosmos—the universe seen as a well-ordered whole.

Because no matter how much it seems that chaos and disorder are inevitable, no matter how impossible it seems to change ourselves, let alone our tiny planet, like the bees we are not prisoners of this destiny. We now know that the environment controls our epigenetics and that we can consciously change our environments. And if enforcing global bans on carcinogenic antibiotics feels daunting, we can start small: whether by choosing a lifestyle that keeps any "cancer genes" we may have been born with safely locked behind their protein coating, or by wrapping a protein coating around a beehive, we can help ensure that a cancer never sees the light of day and that the bees never feel the cold of night. And we should never underestimate our power to do good. Because in spite of all that is dark, it persists. Like consciousness persists. Like bees persist. Like us.

I thought I'd chosen to start beekeeping because I'd witnessed the miracle of the resurrection of a dead bee. And while that was the moment when I made my choice, I've come to realize that what actually defined it was the realization that I had massively underestimated both a man and a bee.

Paul Bond is a highly talented and creative director of photography and an absolutely delightful man. I could forgive myself for thinking that those qualities were more than enough to form an accurate high opinion of him as a human being. But when I found out he was also a world champion beekeeper, I had to add "modesty" to the charge sheet of my crimes of underestimation. When I asked him to tell me the secret behind his global victory, he shyly said, "Oh, I just rinsed out the jars." He explained that his title was judged on the quality of the honey his bees produced and that he'd been lucky enough to have hives in the Chelsea Physic Garden: such was the exotic diversity of delicious plants, his bees almost didn't need to take wing—they could practically crawl into an epic symphony of nectar and pollen.

It's hard to underestimate a living thing more than to assume it's dead. Paul had a hunch that the motionless bee he picked up that cold day might still be alive. What I had underestimated was the effect of the warmth of a helping hand. That it could be so dramatically life-saving. But I now know that the bee Paul had saved was still lost unless it got back to its hive. When we say that honey bees are social insects we don't mean they are merely friendly: they are an integral part of the super-organism that is their colony, and an individual bee cannot survive if it is permanently separated from it.

Bees are flying insects, but as a super-organism the honey bee colony exhibits traits more like those of a mammal: maintaining an internal temperature almost identical to ours; performing essential internal functions by demarcation—individual job categories the equivalent of our internal organs; usually producing only one offspring a year—the super-organism of the swarm.

We are social animals too. Integral parts of larger, social super-organisms. Instances of individual humans being able to survive entirely alone are vanishingly rare. The more we collaborate, the better we live, and to collaborate we need to communicate and connect with each other better: empathy is the master key, as every successful innkeeper knows. It is the only way we can begin to stop fearing people we don't yet understand. Our fellow mortals can seem very different to us—strange and unpredictable, sometimes wildly so. If we don't work to understand each other's very particular, distinctive needs and wants, we cannot successfully live together. Let alone discover that like the honey the bees can spare for us, we too may have all kinds of surpluses that can sweeten and transform other people's lives.

We all need warm helping hands, and we know that we are all capable of offering them. Even those of us who seem to have nothing still find ways to help others. How do we know how to do this? Is it all learned behavior? That we could forget or never be taught? Or is it so important to our survival that some instructions are permanently encoded into our very being? We haven't discovered a gene for human kindness. It may not exist. But if there's even a chance that behaving with kindness, in response to everything that happens to us, might also unlock the tiniest section of protein coating to allow an unmapped, underestimated gene to express itself, let's seize that chance. Let's become humankind. Let's allow each other the possibility of hope.

Why now?

Because the bees show us there is no other time.

Those Two Diary Dates at the Hive

SPRING

When you see pollen being flown into the hive, prepare an empty hive box with waxed top bars, and flip up the overhang of its insulation cover. Cut out a spare top bar cloth. Prepare a honeyed-water spray and your smoker—just in case.

At twilight, when the bees are all tucked up in the hive, flip up the overhang of the insulation of the bottom box on the hive. Remove the heavy roof. Leave the insulated quilt box in place if you (and a friend) are able to easily lift it together with the three inhabited hive boxes—if not, remove the quilt box for the shortest possible time. If the bees have eaten holes in your top bar cloth, add the spare on top.

Lift the three inhabited hive boxes off the base and rest them somewhere safe. Remove the mouse guard and position your prepared empty hive box on the base. Lift the inhabited hive boxes and align them on the new empty box with a star turn.

Flip the insulation overhangs back down, tucking the front of the bottom box's overhang in behind its wool so it doesn't obstruct the hive entrance.

Replace the quilt box and the roof.

AUTUMN

Heft the hive to ensure that the bees can spare you a box of honey.

Gather four small coins, your hive tool and cheese wire, and your Madonna Escape, drip tray, and a strap ready to bind them in place. Load your honeyed-water spray and your smoker. Prepare a new top bar cloth pinned to its applicator, and lay everything out beside the hive as you've rehearsed.

At solar noon, remove the roof, the quilt box, and the insulation from the box you're going to harvest. Nudge the top of the remaining insulation down to clear your work area. With the help of the hive tool and the coins, carefully run the cheese wire through the hive along the bottom of the harvest box. Hurt no bees, and let them settle for a little while.

Peel back the old top bar cloth to let light persuade the queen to move down into the boxes below. Spray down inquisitive bees. Remove the old top bar cloth, replace it with the Madonna Escape with a star turn. Lift the harvest box onto the drip tray with a hovering star turn. Apply the new top bar cloth to the hive, release its applicator, and replace insulation, the quilt, and the roof.

Check that bees leave the harvest box through the Madonna cones.

Fit the mouse guard.

Appendix One

You can buy a Warré hive here: https://www.thewarrestore.com/.

But the supplied hive base is not idle. Here's how to make a replacement that is:

THE IDLE HIVE BASE

In pursuit of thermal equivalence, a very modest four inches of sheepswool have been incorporated into the hive base—emulating a mere fourteen inches of cedar tree beneath our Warré-sized cavity, but of some thermal significance particularly on the many windless days and nights in both winter and summer.

The base is designed around using four-by-two-inch timber in a way that maximizes the accuracy of the specification provided by the sawmill:

The eight pieces of wood in the photo on the previous page all have the same four-by-two cross section as provided by the sawmill: parallel and perpendicular. So resting them on a flat surface next to a vertical wall allows for reasonably precise assembly, without clamps or jigs: screws are more expensive than nails, but minimize plaster damage to the wall from banging!

Note direction of wood grain

The main base frame has the same external dimensions as your Warré hive boxes. The smaller piece of the legs is the height of the four in the four-by-two, the taller just a quarter-inch shy of twice that height. Before fitting the legs to the main base, breathable roofing membrane is positioned on the bottom:

And then breathably clad in an attempt to keep out insects and hibernators:

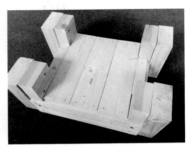

Wool is added inside the main base, which is covered with a standard hive floor:

To prevent any possibility of the insulation on the bottom box sagging and blocking the hive entrance, two "tent poles" are added at the sides of the landing board:

These "tent poles" only need to be tall enough to be flush with the surface of the hive floor, but slightly taller could be useful when nadiring—acting as guides to align the front of the bottom hive box. Space is available on either side of the entrance to fix a mouse guard over winter.

If you can make your own insulated hive base, you can make your own whole idle Warré hive—this was a fundamental premise of Émile Warré's design. Construction drawings are freely available and can be downloaded here: https://warre.biobees.com/plans.htm

THE ROOF

Either get your timber merchant to cut your wood according to the drawings, or do it yourself.

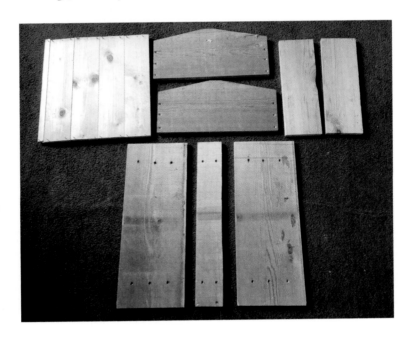

Use your favorite horizontal surface adjacent to a vertical wall to ensure that the base of the roof is level as you screw on the gable ends:

I use tongue-and-groove slats to allow the wool in the quilt box below to breathe, without allowing moths access.

And then you're ready for the pitch and the apex:

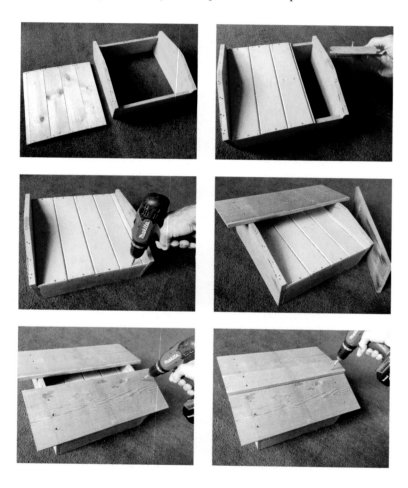

THE HIVE BOX

Top bars need rebates to sit in the two shorter pieces of wood you've cut in accordance with the construction drawings. If your timber merchant can cut these rebates on his table saw, this is ideal. If you're doing it yourself, you need to make a guide to help you saw the rebates reasonably accurately. Cut three lengths of half-inch thick timber longer than your short side and glue two of them together to make a handsaw guide:

Rest your short side up against that horizontal/vertical kitchen worktop/wall with the half-inch timber positioned like this, and use it to guide your saw as you hold it firmly in place while protecting your hand by distancing it from a slip of the saw with a hammer or equivalent:

Carefully saw down to slightly more than the height of the top bar the rebate's going to accommodate.

Then lie your short side flat, draw a line just below the height of the top bar it's going to accommodate, and use the saw guide to saw down to meet the cut you've just made. It's important to saw no deeper than this: doing so will unnecessarily weaken the load-bearing top edge.

Assemble the box so it's as level as possible:

Carefully draw a horizontal line along the middle of the long side and fix the handles centrally so their bottoms align with it.

Nail triangular moldings of wood to the underside of your top bars to give the bees a straight edge to hang comb from, making sure you leave room for the ends of the top bar to fit into the rebate:

Appendix Two

To insulate your idle Warré hive:

Buy the imminently available bespoke product, or if you can't wait:

Source clean sheep wool, free from pesticides. I get mine from a company that makes woolen insulating packaging used by organic food suppliers to keep their products fresh in transit. It's also used to keep human organs and vaccines chilled in transit. It came on a roll whose width turned out to be exactly six times the height of a Warré box.

Cut three strips long enough to comfortably wrap around your Warré box, and fold over to make double thickness. Cut any cheap fabric to make a cover, like a pillow slip, for the wool strips.

I used a sewing machine to make easily removable sleeves, but if sewing is not your forte, you can use a staple gun to more permanently fix everything to the hive boxes. The principles are the same.

Sew the fabric into a sleeve, and stitch the hem to allow a sisal drawstring to be inserted, and move freely, the length of the sleeve.

Place an identical length of sisal inside the fold of the strip of wool when you fold it over double, and insert the wool into the sleeve. Sew the ends of the sleeve closed, ensuring that both lengths of sisal can slide freely and act as drawstrings:

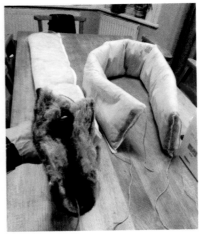

(Hold the fabric in place with glue if you're stapling it onto the hive box—no drawstrings required.)

Attach breathable roofing membrane as weatherproofer. http://www.protectmembranes.com/uploads/ddd45e1cb3584a908c2ff b62079b4bb1.pdf

Cut strips thirteen inches wide (to allow an overhang) and ten inches longer than the woolen sleeves. Using a needle strong enough for denim, sew the membrane to the hem of the sleeve with five inches extending either end:

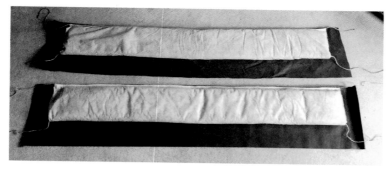

Stop the stitching short of the ends of the sleeve by three inches to allow for more convenient tying of the drawstrings:

(Just glue the roofing membrane in place if you're stapling.)

Repurpose office materials to attach the sheepswool sleeve to a Warré box, and seal the roofing membrane:

Each hive box needs two large paper clips and one plastic slide binder. Trim the slide binder with a hacksaw so it's a little shorter than the height of a hive box—and preserve the end with the curved guide. The bulldog grip is an essential third hand!

Tie the woolen sleeve in place with the two drawstrings:

Use the bulldog grip to hold one end of the woolen sleeve onto the corner of a box while the sleeve is tightly wrapped around. Tie the drawstrings together:

The top drawstring needs to be as near to the top of the hive box as possible. A good way to achieve this is to tie it tightly in place with one corner of the box slightly covered—and then use your thumbs to prize the drawstring back down onto the vertical corner edge: it's hard work, but the drawstring becomes very tight indeed, allows for precise manipulation, and will stay where you put it!

The bottom drawstring should similarly be as near as possible to the bottom of the hive box.

To seal the roofing membrane:

Align the two ends of the membrane and carefully hold with the bulldog grip:

Slide the slide binder down through the bulldog to grip the ends together:

 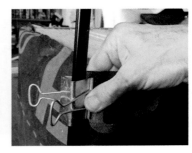

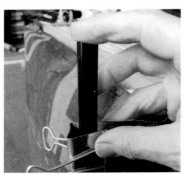 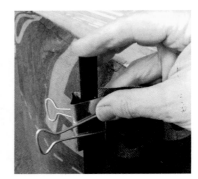

Wind the slide binder toward the hive box making a tight roll of membrane:

Hold the roll in place by paper clips—at the top of the membrane insert the small loop of a paper clip inside the slide binder in the middle of the roll, the large loop between the roll and the woolen sleeve:

(If you're stapling you can leave out the slide binder, roll the two ends together tightly by hand, and staple to the top of the box.)

Making sure the roll at the bottom is as tight as possible (it's beyond the full influence of the slide binder, which is too short to reach the bottom), insert the large loop of a paper clip into the center of the roll and the small loop between the membrane and the hive box:

This hive box is now insulated, and the sealed roofing membrane that overhangs the box below can keep the insulation wind- and water-proof.

To remove this box from the hive, simply extract the bottom paper clip, loosen the roll of overhang, and fold it up (the paper clip in the photo is just making sure it doesn't get lost!). The slide binder ensures that the roofing membrane covering the insulation stays in place, and only the overhanging membrane is flipped up:

The insulation drawstrings can be nudged up and down at the corners to expose the bottom of the hive box:

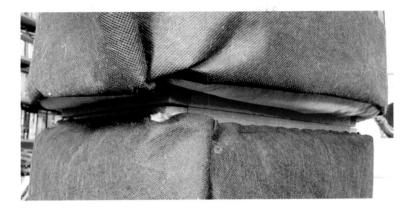

The first time you perform this task it might take all of a minute, but subsequently it's difficult to make it last more than thirty seconds!

The hive box can now be treated in the conventional way and lifted by the handles gripped through the insulation. The top and the bottom surfaces of the box are unencumbered by the insulating process:

After replacing the box, edge the insulation drawstrings up and down to where the hive boxes meet, and fold down the membrane over-hang. Reconfigure the bottom of the rolled join in the membrane and reinsert the paper clip. The process will come to take less than a min-ute with practice.

The insulating component of this hive box could now be consid-ered as integral to it as its top bars: put in place before that box is nadired into the hive, and then, with luck, left alone until the box rises to be harvested some years later, when any need for maintenance can be assessed. The insulation, membrane, and paper-clip fastening on my boxes have survived three winters and are still going strong.

THE QUILT BOX

In an attempt to emulate the thermal resistance of what might be as much as twenty feet of tree trunk above the wild nesting colony living in the Warré-sized cavity in our hypothetical cedar tree, the height of the quilt box was doubled from four to eight inches—the size of a hive box.

Twenty feet of cedar has a thermal resistance of R336. Eight inches of wool is R28—the equivalent of twenty inches of cedar above our wild nesting cavity. Not a flagrantly excessive amount of tree to be emulating.

So while some might feel eight inches of wool in the quilt box is only needed during the kind of winters bees encounter in Siberia, it's worth noting that even in the temperate Soft South of the United Kingdom, current building regulations stipulate a minimum of eight inches loft insulation for human dwellings—in suburban homes clad with wisteria, eight inches is not thermal hysteria! They'd want more in Siberia!

Here's how the eight-inch quilt box connects the Warré roof to the top hive box: a burlap bottom and four simple swivel stays are added to a hive box that's filled with sheepswool:

The stays do not need to waterproof the join between quilt and hive box—they're just stopping it sliding off.

Back to the sewing machine (or glue pot) for a slightly different woolen sleeve: same length as before, but this time the wool is a single thickness and only four inches wide:

There's a very wide hem of a good half-inch, but no drawstrings.

The roofing membrane waterproofing is the same length as before but narrower—about ten inches—and this time the sleeve is stitched right up to its ends, but about half an inch from the top of the membrane:

 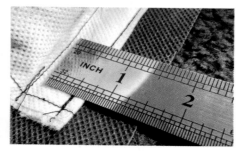

Fit the Warré roof onto the eight-inch quilt box and draw a line around its bottom edge, then draw another line 1.5 inches above it:

With a thumbtack, pin the middle of the quilt woolen sleeve to the middle of the front of the quilt box, so the top of the membrane follows the top line:

Move onto a side and pull the membrane as tight as you can while still following the line with the top edge: because of the thickness of the insulation, this will create a pleat/fold in the membrane at the corner. This is fine, just keep it to a minimum. Pin the membrane in place on the line in the middle of the side.

Repeat with the other side, and at the back:

Pin down the pleats/folds and check that the two ends of the membrane align:

Once you're satisfied it's tight and following your line, roll the two ends of the membrane together as tightly as you can toward the quilt box and staple the roll to the quilt box. Then staple the rest of the top of the membrane to the line and remove the thumbtacks:

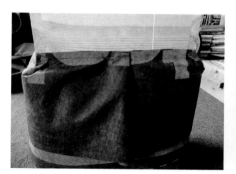 

Make the bottom of the membrane roll as tight as possible and fix with a paper clip as before:

The roof overhangs the top of the quilt's roofing membrane:

To remove the quilt box, the bottom paper clip is taken out and the membrane and insulation folded up. The stays can then be rotated to the unlocked position and the quilt box lifted away:

When replacing the quilt box, after locking the stays, fold down the woolen sleeve first and make sure it's properly in place with the two ends meeting squarely, before folding down the membrane, reinstating the roll, and reinserting the paper clip—it won't take a minute.

Down at the base:

The overhang at the front of the bottom box must be tucked up under the woolen sleeve so it can be propped up by its "tent poles":

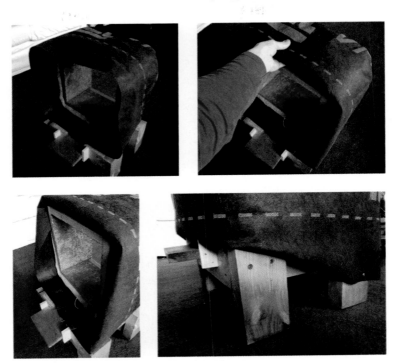

The walls of this Warré hive now emulate the thermal resistance of our hypothetical cedar tree, but they're not identical: the layer of wool cannot carry the thermal mass of eight inches of cedar.

But the cedar never naturally evolved to be an insulator, whereas the wool spent about eight million years evolving to help keep an organism alive at a constant temperature of 100 degrees Fahrenheit. We are now discovering all sorts of clever ways wool does this—much of the significance of the breathability of the materials covering the wool in this hive insulation system is to allow the wool to proactively respond to the changing atmospheric humidity.

And though 100 degrees Fahrenheit isn't perfect for the 95 degrees that the bees need to maintain for their young, it's pretty close . . .

CALCULATIONS

Cedar has a Thermal Resistance of R1.4 per inch: http://energy.gov/
energysaver/articles/energy-efficiency-log-homes

Wool's is R3.5 per inch:
http://energy.gov/energysaver/articles/insulation-materials
Professor Tom Seeley found wild nesting hives' Thermal Resistance
ranges from an average R6 to a cozy R29.

Two layers of wool = 2.75 inches × R3.5 = R9.6
(In addition to R1.4 of one-inch cedar)
Total Thermal Resistance of wool-clad box = R11

To achieve the R9.6 added by the wool, 9.6 divided by 1.4 inches of
additional cedar would be required = 7 inches. So to achieve the same
Thermal Resistance of R11, the solid cedar hive walls would need to
be eight inches thick.

Weight of R11.2 Hive Box:

Cross-sectional area
of one-inch-thick box =
$(14 × 14) - (12 × 12) = 52$
Cross-sectional area of
eight-inch-thick box =
$(28 × 28) - (12 × 12) = 640$

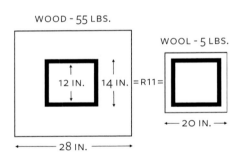

WOOD - 55 LBS.

WOOL - 5 LBS.

12 IN.   14 IN. =R11=

20 IN.

28 IN.

The eight-inch-thick cedar box is heavier by a factor of 12.3: 55 pounds
empty, potentially 100 pounds full of honey
    The thermally equivalent one-inch-thick cedar box wrapped in
wool weighs 5 pounds.

Appendix Three

MAKING A CONVERTER BOX

See the video of a converter box in action at https://www.youtube.com/watch?v=nAzL_VPGgNs

When you buy a nucleus of bees, it will consist of a microcosm of a colony living on five frames of honeycomb. These are too long to fit inside a Warré hive box, so we have to provide temporary accommodation that will allow these framed bees to build their home downward into your Warré hive. We're going to make a box like the one your bees have just arrived in, but without a bottom, and we're going to enclose its position on top of the Warré to make your bees safe and snug.

Buy a flat pack Nucleus box and assemble it according to the manufacturer's instructions, BUT LEAVE OUT THE BOTTOM.

Make sure any ventilation holes or entrances that have been provided are filled:

Cut four lengths of three-by-one-inch timber:

- Two longer lengths of roughly nineteen inches (the external width of your Warré hive box plus two widths of the three-by-one-inch timber)
- Two shorter lengths of roughly fifteen inches (the width of the shorter side of your nucleus box plus two widths of the three-by-one-inch timber)

Center the longer lengths of three-by-one flush with the bottom of the longer sides of the nucleus box:

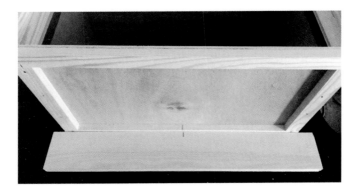

Fix using glue and nails hammered from the inside. Tip the nucleus box on its side:

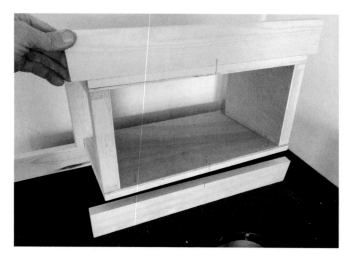

(Use the shorter lengths as temporary props to hold the side level while you hammer.)

Turn the nucleus box upside down and align your Warré hive box centrally:

Align the shorter lengths of three-by-one on top of the longer so they rest up against the side of the Warré hive box and screw in place:

Turn everything the right way up and it should look like this:

When your bees arrive, you can transfer the frames they're living in—it's essential to keep them in the same order relative to the hive entrance.

Line up the frames so they're directly above your Warré top bars—to keep the descending comb architecture and bee space as uninterrupted as possible. To fine-tune this, you can slightly adjust the position of the nucleus box by sliding it to and fro on the Warré hive box if need be.

Cut a rectangle of top bar cloth to fit on top of your nucleus box, make a rectangular quilt box the way you make a Warré hive box (see appendix 1) but to the same external dimensions of the nucleus box, fill it with sheepswool, and the roof supplied with your nucleus box should neatly fit on top:

INSULATING AND WEATHERPROOFING
THE CONVERTER BOX

Our standard Warré hive box insulation sleeve (see appendix 2) *almost* fits the nucleus box: the woolen part is sufficient, but it needs longer breathable roofing membrane to accommodate your additional joinery—an extra three inches at both ends will do it.

Again, the quilt box sleeve in appendix 2 has enough wool to go around, but here the roofing membrane needs to be extended up higher so it can fit under the nucleus box roof's overhang—an extra four inches should be plenty.

Appendix Four

THE MADONNA ESCAPE

Make two very simple square frames from one-by-one-inch timber with the same external dimensions as a Warré hive box. Staple squares of Correx to fit the frames, as shown below.

Take a piece of paper and roll it into the shape and size of cone you're aiming for—don't worry about precision at this stage, you're just making a quick template. Make the apex pointier and the base bigger than you need. Tape down the corner of the paper on the outside to hold your cone in place. Trim the base end of your paper cone to create a roughly circular base. Then make one half inch cuts in the base every half inch so you can fold up tabs all around and your cone can now stand up on its own base like a traffic cone.

One of your frames is the drip tray. On the other frame cut four holes in the Correx that are the same diameter as the base of your paper traffic cone.

Carefully remove the tape, unroll the paper cone, and now use it as a template to cut four identical shapes of aluminum fly screen. This material is just like thick kitchen foil with holes, very malleable, and can be cut with ordinary scissors that aren't too precious. Form your cut fly screen into clones of your paper cone: where you taped to hold the paper, now use a short length of thin wire to fix that corner of the screen.

Push the cone through the hole from underneath and glue the tabs to the inside of the Correx. There are many pens that conveniently have a 0.35 inch diameter girth—big enough to allow even a fat drone to squeeze out, but small enough to prevent any bee returning. Scrunch the fly screen at the apex to hug the pen and you're done.

Appendix Five

The cashmere cluster experiment: to create a very rough approximation of a winter "cluster" to compare energy used to maintain its temperature during the winter by an insulated (idle) and an uninsulated Warré hive.

An 18-watt lightbulb was the heat energy source. A digital thermostat controlled the internal "cluster" volume temperature at 25 degrees centigrade (77 degrees Fahrenheit) until the beginning of February when the brood would require the temperature to be raised to 33 degrees centigrade (92 degrees Fahrenheit). A digital meter measured the electrical energy consumed by the 18-watt bulb in watt hours.

Each hive consisted of three Warré boxes. The bulb (controlled by the thermostat sensor just above it) was suspended from the top bars:

The bulb and sensor were both completely enclosed in a soccer ball-sized sphere covered with cashmere wool (from a sweater previously loved by moths!).

The two "clusters" in place:

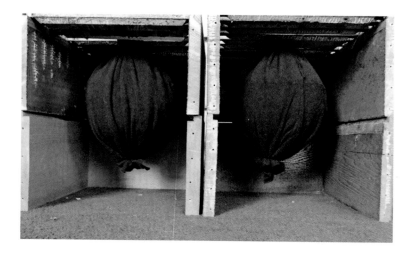

This passive wool-lined "cluster" bears very little resemblance to the complex, dynamic heating/insulating effect of a cluster of bees in honeycomb (bees are both heaters and insulators; honey is both fuel and a thermal mass of stored heat that diminishes as it is consumed). But at the very least it delineated a boundary for the "cluster" in the experiment—the bulb is having to do less immediate work heating up the hive beyond the wool boundary: the thermostat is controlling for 25/33 degrees centigrade (77/92 degrees Fahrenheit) within the wool, not for the whole hive cavity. A small step toward an emulation of the bees' heating work, and easily replicable in both the insulated and uninsulated hive.

(Humans with a body temperature of 98 degrees Fahrenheit have some idea of the insulating effect of wearing a cashmere jumper!)

So with identical "clusters" and cavity sizes, one hive was weatherproofed with breathable roofing membrane and insulated with Woolcool sheep wool to an equivalent of R11 in its walls, R28 in its eight-inch quilt box, and R14 in its four-inch base. The other, traditional hive was comprised of R1 uninsulated three-quarter-inch cedar

hive boxes and base, and its four-inch quilt box was filled with straw
to give R5. Both were assembled with a single cable running to the
control box:

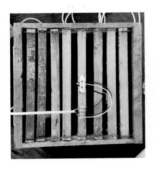

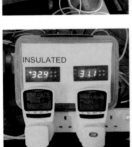

The amount of energy (watt hours) each hive consumed to keep the
thermostat inside the wool-lined "cluster" at 25/33 degrees centigrade
(77/92 degrees Fahrenheit) was measured, along with indicative ambi-
ent maximum and minimum temperatures for this period:

DATE (18W BULBS) STAT @ 25C	AMB. MIN °C	AMB. MAX °C	INS. W HOURS	UNINS. W HOURS	INS. DAILY W HOURS	UNINS. DAILY W HOURS	UNINS. EXTRA %
2018 Jan 1	7	13	171	256	103	156	151.5%
2	5	7	284	421	113	165	146.0%
3	3	12	392	578	108	157	145.4%
4	4	11	503	739	111	161	145.0%
5	3	11	616	904	113	165	146.0%
6	3	8	736	1080	120	176	146.7%
7	2	6	875	1284	139	184	132.4%
8	2	6	1011	1481	136	197	144.9%
9	3	6	1137	1668	126	186	147.6%
10	4	10	1250	1831	113	163	144.2%
11	4	10	1358	1991	108	160	148.1%
12	4	8	1470	2156	112	165	147.3%
13	5	9	1587	2327	117	171	146.2%
14	4	8	1705	2500	118	173	146.6%
15	3	8	1820	2664	115	164	142.6%
16	5	12	1927	2823	107	159	148.6%
17	2	7	2055	3011	128	188	146.9%
18	2	12	2176	3184	121	173	143.0%
19	3	9	2296	3360	120	176	146.7%
20	1	6	2427	3551	131	191	145.8%
21	3	5	2561	3753	134	202	150.7%
22	1	10	2674	3917	113	164	145.1%
23	7	11	2778	4069	104	152	146.2%
24	10	14	2866	4197	88	128	145.5%
25	7	14	2968	4349	102	152	149.0%
26	5	10	3082	4515	114	166	145.6%
27	4	9	3205	4694	123	179	145.5%
28	5	11	3306	4841	101	147	145.5%
29	6	14	3392	4965	86	124	144.2%
30	3	13	3495	5119	103	154	149.5%
31	0	7	3609	5281	114	162	142.1%
2018 Feb 1	3	8	3730	5463	121	182	150.4%
2	4	8	3859	5650	129	193	149.6%
3	2	8	3987	5835	128	185	144.5%
4	2	4	4114	6025	127	190	149.6%

DATE (18W BULBS) STAT @ 25C	AMB. MIN °C	AMB. MAX °C	INS. W HOURS	UNINS. W HOURS	INS. DAILY W HOURS	UNINS. DAILY W HOURS	UNINS. EXTRA %
5	1	5	4245	6220	131	195	148.9%
6	2	6	4384	6424	139	204	146.8%
7	0	5	4528	6637	144	213	147.9%
8	-1	6	4669	6843	141	203	144.0%
9	0	8	4791	7021	122	178	145.9%
10	-1	6	4932	7230	141	199	141.1%
11	0	10	5051	7407	139	177	127.3%
12	0	8	5185	7606	134	199	148.5%
13	0	7	5304	7777	119	171	143.7%
14	-1	6	5427	7963	123	171	139.0%
15	2	8	5545	8135	118	172	145.8%
16	1	10	5669	8316	124	181	146.0%
Stat 33C 17	1	9	179	261	179	261	145.8%
18	2	10	349	512	170	251	147.6%
19	5	11	497	731	148	219	148.0%
20	7	11	645	955	148	224	151.4%
21	3	9	814	1206	169	251	148.5%
22	3	8	996	1481	182	275	151.1%
23	1	6	1178	1756	184	275	149.5%
24	0	5	1378	2059	200	303	151.5%
25	0	5	1564	2342	186	283	152.2%
26	0	5	1762	2642	198	300	151.5%
27	-3	0	1974	2968	212	326	153.8%
28	-4	-1	2188	3296	214	328	153.3%
2018 Mar 1	-5	-1	2415	3644	227	348	153.3%
2	-4	0	2634	3981	219	337	153.9%
3	-3	1	2849	4313	215	332	154.4%
4	2	4	3036	4601	187	288	154.0%
5	5	9	3173	4809	137	208	151.8%
6	5	11	3327	5041	154	232	150.6%
7	5	11	3480	5274	153	233	152.3%
8	5	10	3641	5517	161	241	149.7%
9	4	10	3805	5763	164	246	150.0%
10	6	11	3955	5991	150	228	152.0%
11	8	15	4098	6201	143	210	146.9%

At the height of the summer of 2018, I removed the mouse guards and the cashmere so that the bulbs had to heat the whole of both hives. In the first two weeks of July when temperatures reached 30 degrees centigrade (86 degrees Fahrenheit), although both hives used less energy than in the winter, the uninsulated hive still used more than 50 percent more.

DATE STAT @ 33C	AMB. MIN °C	AMB. MAX °C	INS. W HOURS	UNINS. W HOURS	INS. DAILY W HOURS	UNINS. DAILY W HOURS	UNINS. EXTRA %
2018 July 2	14	28	108	163	108	163	150.90%
3	14	26	222	363	114	200	175.40%
4	12	27	343	563	121	200	165.30%
5	16	28	449	744	106	181	170.80%
6	17	30	538	882	89	138	155.10%
7	17	30	626	1016	88	134	152.30%
8	17	30	713	1149	87	133	152.90%
9	15	29	817	1320	104	171	164.40%
10	14	23	923	1503	106	183	181.30%
11	14	24	1054	1735	131	232	177.10%
12	16	25	1180	1950	126	215	170.60%

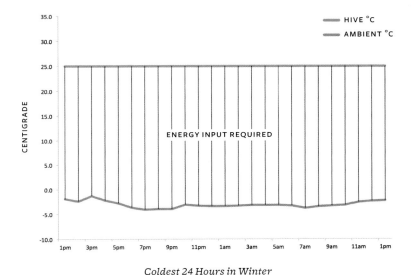

Coldest 24 Hours in Winter

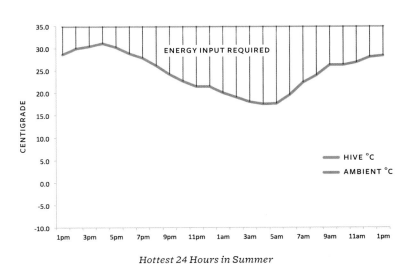

Hottest 24 Hours in Summer

Acknowledgments

Matt Casbourne turned a pitch into a book. David Heaf, Nicola Bradbear, and Joe Walker generously infused the manuscript with their rigorous wisdom. The encouragement they all gave was beyond measure.

BILL ANDERSON is an urban beekeeper and educator based in London who writes the regular beekeeping column for *The Idler* magazine. His online Idle Beekeeping course is currently available from The Idler website. In the other 363 days he isn't tending to his hives, Anderson is a television drama director, working on a huge variety of shows, including *Dr. Who* and *Mr. Selfridge*.